ISBN 978-0-692-12988-3

Thinking Like A Scientist

Using Simple Math to Understand Science and its Applications to Everyday Life

i

Acknowledgments

I am hugely indebted to Tom, Colin, Barbara, Kathleen, and John, thoughtful and careful readers one and all.

Preface

What does it mean to be a well-educated person? The definition depends on time and culture, and what passed for a good education one or two generations ago is hopelessly out of date today. To be well educated means keeping up with the times through self-education, through newspapers and magazines and television and the internet. Formal education, no matter how good, goes stale in a hurry. A well-educated person reads, thinks, and stays abreast of his or her times, and has a fair understanding of the workings of society and the world beyond. If knowledge of Shakespeare and the Bible were once sufficient reading in the English-speaking world to understand human nature, that has long since ceased to be true.

No one can have enough education to understand everything that is important. Every organization is run by specialists. Doctors and lawyers and teachers and sailors and pilots are all specialists; so are police and firefighters. Most people over the age of 21 are specialists of one sort or another. A well-educated person is supposed to understand, in a vague way, how each of these specialties works, and how to behave when interacting with someone whose business is something of a mystery. All of us need to understand money and the stock market or find someone who is trustworthy and skillful to do our thinking and investing for us. We need to understand enough about medicine to choose a good doctor – or find someone who is trustworthy and skillful to find one for us. Lawyers, teachers, bankers, contractors, politicians, trainers and tax preparers – an educated person needs to know enough about each specialty to make wise choices, or else know someone who can help us choose, or who can choose for us. A well-educated person knows enough about how the world works to make good decisions, but how can I make a wise choice if I know little or nothing about the business at hand? Do I really have to know something about a subject in order to have an intelligent opinion about it?

Those who manage large organizations are often faced with this problem. How do I make decisions when faced with conflicting advice? There are ways of probing. Who seems honest, who seems self-serving? Who seems knowledgeable, who seems incapable of clear explication? Who is experienced, and who is a novice? Such judgments usually fall under the rubric of "character:" Whose character can I trust? I am reminded that a very well-educated American wartime President was widely regarded by his closest advisors as having a "second-rate intellect, first-rate temperament." If you have a first-rate temperament or are a great judge of character, you can probably make good decisions and choices without knowing much about what you are doing. If not, perhaps you'd better work harder at being well educated.

So what does a well-educated person, living at the beginning of the twenty-first century, need to know? There's reading and writing and arithmetic, of course, and history and economics and political science. You need to know how to use a computer and how to drive a car. A little math is useful, but I'm told in real life there isn't any algebra, although you may need it in your profession. Given the proliferation of take-out, cooking seems a skill that can be dispensed with, although the economic cost of not knowing how to cook is probably on a par with not knowing how to drive. But these skills, although essential to being a generally competent person, have little to do with the core sense of "well educated," which has usually been understood largely in the context of citizenship and politics. A well-educated person understands the political issues of the day, and votes knowledgeably, if not necessarily intelligently. The key issue is knowledge: What does a well-educated person have to know to be a good citizen?

I do not believe that voting and citizenship are areas that can be turned over to expert advisors who vote our interests as they see fit. (This is the corporate governance model.) As the founders of American democracy foresaw, "factions" vote self-interest, which is often narrowly and unwisely defined, and factionalism does not reward men and women of character. There is no substitute for the well-educated citizen.

So what does it take to be well-educated?

Well, I think some familiarity with mathematics and science should be part of your toolkit. This book – and, let's face it, this is a TEXTBOOK, with equations and problem sets – tries to show how critical thinking, coupled with a little science and math, can shed light on matters that are of general public concern, from public health to tides and eclipses, from cell phones to x-rays.

Why would anyone want to read a TEXTBOOK that hasn't been assigned in class for a grade and a degree? The answer depends on your level of curiosity. This book illustrates how scientists apply logic and math and physics and chemistry and biology and astronomy to approach topics outside their immediate expertise. Few people are experts at more than a few subjects – it is always wise to be suspicious of those who, because of their success in one field, claim expertise in others – but anyone equipped with a little background in science and math can go on to get a better understanding of public issues. We live in a politicized world, a technological world. Knowing how to think about science has become a survival skill, a means of uncovering frauds and grasping opportunities. Like vegetables, it's good for you, and like fruits, it can be tasty. And fun.

A Note on Organization

The first chapter of this book is labeled Chapter 0. **Don't read it! Not now!** If you were educated outside the United States of America, you can skip it entirely. Or skim it. If you have the misfortune to have had an American education, you'd better read it at least once and probably several times. Not all at once, of course, and not right away, but use Chapter 0 for reference. The subject, of course, is the METRIC SYSTEM, used by essentially all nations except the USA. It's a wonderful, logical, thoughtful, and careful system of 'weights and measures,' encompassing length and area and volume and density and energy and power and temperature and pressure and pretty much everything a scientist would like to measure and talk about. Scientists, even American scientists, use the metric system without exception.

The rest of the book is divided loosely into four sections: Things that Change with Time; Health and Risk; Energy and the Future; and Scientific Tools and Methods. Every section is largely independent of every other; every chapter is pretty much independent of every other. Unlike a regular textbook, you the reader can pick and choose as you will. **But have a glance at Chapter 0. And don't let it scare you off! Treat this chapter like an appendix, a reference. It's dense and** heavy, hard work for most readers. Come back to it occasionally. But start this book somewhere else!

Thinking Like A Scientist

Table of Contents

Part III: Energy and the Future

Part IV: Scientific Tools and Models

Chapter 0: Weights and Measures

Hey Reader! Did you see what I said in the Preface? Start this book somewhere else!

 The units of measure that scientists use are a patchwork mix of old definitions chosen for convenience and new units chosen for consistency. When atmospheric pressure is reported as "30.05 inches and rising," that's old-fashioned convenience speaking, dating from the days when a barometer was a glass tube of mercury standing upright in a mercury pool (Figure 0.1). When energy is reported in Joules, that's modern consistency: A Joule is the energy delivered in one second by a current of one ampere at a potential difference of one volt (E = IVt), and also the energy in a two kilogram mass moving at one meter per second (E = ½ mv^2). Convenience and consistency both have their virtues and faults, but the resulting mix is often bewildering and infuriating. In this book I have used whatever units are customary, so that, for example, energy is measured in Joules, kilowatt-hours, BTUs, barrels of oil, and cubic feet of gas. In most cases I have made an effort to provide a conversion to the common modern unit (Joules), but this does not always eliminate the sense of bewilderment.

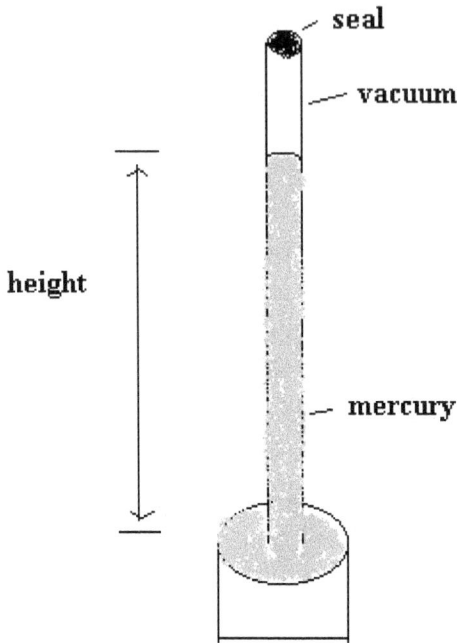

Figure 0.1 The Torricelli Barometer

If you live in the United States, confusion is part of your heritage. You have to remember that there are 12 inches in a foot, three feet in a yard, 5280 feet in a mile, and 640 acres in a square mile; sixteen ounces in a pint, two pints in a quart, four quarts in a gallon (yes, that's 128 ounces in a gallon); a pint of water weighs a pound, so a gallon of water weighs eight pounds; the freezing point of water is 32° F, the boiling point is 212° F. And let's not forget that a barrel of oil contains 42 gallons, although I think most of us **can** forget that a bushel is 32 quarts (or 4 pecks). How many gallons in a cubic foot? The answer is 7.481, and don't ask why. Neither a teaspoonful nor a tablespoonful is found in my handbook, but my cookbook says there are three teaspoonsful in a tablespoon and sixteen tablespoonsful in a cup. Oh yes, a cup is eight ounces, half a pint. By the way, a pint of water, 16 ounces, actually weighs a bit more than a pound, so a gallon of water weighs 8.337 pounds (but a British gallon weighs 10 pounds). The United States is the only developed nation on earth that has not abandoned its traditional system of weights and measures, and the reason is simple enough: The U.S. economy is so large that there is no pressure, internal or external, to change. Manufacturers, of course, routinely use the international metric system if they want to sell products elsewhere in the world, and U.S. law permits, but does not require, metrification. This approach helps commerce but puts an extra burden on anyone wanting to think like a scientist.

Scientists use the metric system. It's so much less work. The metric system or SI (Systeme Internationale, in French) has only three basic units of measure: The second (for time), the meter (for distance), and the kilogram (for mass or weight). Nearly every other measure in the metric system derives from these three.

Let's start with the second. It is the only unit not based on powers of ten. Sixty seconds to the minute, sixty minutes to the hour, twenty-four hours to the day, and 365.24 days to the year. Not a simple system, not an obvious system, its only virtue is that it is universally learned and understood by every educated (or uneducated) person around the globe. In a system of weights and measures, universality is far more important than simplicity, although I confess that I wish the French revolutionaries who invented the metric system had succeeded in decimalizing the day. Too bad.

The meter was chosen so that the distance between the North Pole and the Equator was ten million meters, or ten thousand kilometers. A Good Idea, but the earth is not quite a sphere, so the circumference of the earth is not exactly forty million meters. But it's close enough, for most purposes. Like the second, the meter is an arbitrary human construct, but is, at the very least, roughly based on a property of the planet we all inhabit. A centimeter is one one-hundredth of a meter; a kilometer is a thousand meters. Simple enough. And, in U.S. law, an inch is defined as exactly 2.54 centimeters, from which the entire U.S. measurement system for length and area and volume can be related to metric counterparts. A foot is (exactly!) 12 x 2.54 centimeters (30.48 cm), and a mile is 160934.4 centimeters, or 1609.344 meters, or 1.609344 kilometers.

The kilogram was chosen so that the density of water is (very nearly!) one gram per cubic centimeter, or if you will, one thousand kilograms per cubic meter. A thousand kilograms is called a tonne, or, in English, a long ton, two thousand two hundred pounds, which distinguishes it from the short ton, two thousand pounds. Thus, a kilogram is 2.2 pounds, and a cubic meter of water weighs a tonne. Unfortunately, U.S. law does not define the pound in this way. Legally, a kilogram is 2.2046223 pounds, which is not something you need to know. As far as we are concerned here, a kilogram is 2.2 pounds. Furthermore, the density of water depends on temperature, so a cubic meter of water does not weigh exactly a tonne at all temperatures. But it's close enough.

From the meter we obtain measures of the area (square meters, or m^2) and volume (m^3); from the meter and second, we obtain measures of speed (meters per second, or m/s) and acceleration (meters per second squared, or m/s^2). From the meter, kilogram, and second taken together we obtain all the other important units, such as energy, force, and pressure. We will get to these in a moment. First, we have to try to clear up a small discrepancy concerning the liter. A liter of water weighs a kilogram. A liter is by definition the same volume as 1000 milliliters (ml). Thus, a cubic centimeter (cc) and a milliliter (ml) ought to be identical. Legally speaking, they are not. Practically speaking, they are. There is a small discrepancy in the legal description, but it is of no practical importance. A liter is a measure of volume, 33.8147 ounces or 1.05671 quarts, but in practical terms there are 1000 liters in a cubic meter. For all practical purposes, a liter of water weighs a kilogram.

Next, we consider the four most important units derived from the meter-kilogram-second scheme: The Joule, a unit of energy; the Watt, a unit of power; the Newton, a unit of force; and the Bar, a unit of pressure. We start with the Joule.

The energy contained in a moving object is $E = \frac{1}{2} mv^2$, where m is the mass of the object and v its velocity (speed, if it is moving in a straight line). Thus, an object with mass 1000 kg (kilograms) moving at a speed of 10 meters per second has energy 50,000 Joules. More to the point, the basic unit of energy, the Joule, has dimensions $(kg)(m/s)^2$.

Power is energy (lost or gained) per second. The basic unit of power is the Watt, which has dimensions $(kg)m^2/s^3$. A Watt is a Joule per second. One of the beauties of the international metric system – indeed, its principal beauty – is that the Watt defined by mass, distance, and time is the same Watt defined by electrical voltage and current. A Watt is also a current of one ampere passing through a voltage drop of one volt.

Force is mass times acceleration, so its units are $(kg)m/s^2$. A kilogram of mass accelerated at one meter per square second is experiencing a force of one Newton. Because of gravitational attraction, every object on the surface of the earth experiences a gravitational force proportional to its mass. We call this force "weight," and the proportionality between weight and mass is known as the local gravitational acceleration. On the surface of the earth near sea level, this acceleration, known as **g**, is 9.81 m/s^2, or, in the U.S. system, 32 feet per second per second. Thus, if your mass is 75 kilograms, your weight is 736 Newtons. On the moon, your mass would be the same, but your weight would be far less; in a spacecraft orbiting the earth, you would be weightless.

Energy is also gained or lost when an object is raised or lowered against the force of gravity. If we raise a one-kilogram mass by one meter in height, we add 9.81 Joules to its gravitational potential energy. The formula is (kg)mg, that is, mass (kg) times height (m) times the acceleration of gravity (**g**). The reader should confirm that the units are correct, that is, $(kg)m^2/s^2$.

Pressure is force per unit area, so its units are $(kg)/ms^2$, and the basic metric units of pressure are the Bar and the Pascal. Consider the pressure of the atmosphere on the surface of the earth. Standard atmospheric pressure is a column of mercury 760 millimeters in height, or 0.76 meters (29.92 inches in the U.S. system). The density of mercury is 13595 kilograms per

7

cubic meter (13.595 grams per cc), so the pressure exerted by a column of mercury 0.76 meters high is 9.81 x 0.76 x 13595 = 101,325 kilograms per meter per square second. By convention, one Bar is 100,000 (10^5) kilograms per meter per square second, so normal atmospheric pressure is 1.013 Bars or 1013 milliBar.

The pressure unit known as the Pascal is more consistent with the spirit of the metric system: A Pascal is defined as one Newton per square meter, while a Bar is 10^5 Newtons per square meter. But the fact that normal atmospheric pressure is very nearly a power of ten metric multiple – a special and accidental convenience -- makes the choice of the Bar irresistible. But scientists use both the Bar and the Pascal as basic units of pressure; all you have to remember is that a Bar is 10^5 Pascals. The kiloPascal (kPa) is also widely used. A Bar is 100 kPa.

Energy comes in a variety of forms: Kinetic (energy of motion), gravitational (energy of height), thermal (energy of heat), electrical, nuclear, chemical, and so on. All forms of energy can be converted into heat, so for a long time measures of thermal energy were used as the common currency. A calorie, for example, is the energy required to heat a gram of water by 1°C; a kilocalorie is the energy required to heat a kilogram of water by 1°C. (The food Calorie with which we are all familiar is actually a kilocalorie; it should always be written with an upper case

'C' to distinguish it from the calorie.) A BTU, or British Thermal Unit, is the energy needed to heat a pound of water by 1°F. These are wonderfully convenient units for experimental measurements of energy – all you need is some water and a thermometer – but they are inconsistent with the rest of the metric system. Unfortunately, the conversions among different systems of energy units are hard to remember. A calorie is 4.186 Joules, so a kilocalorie is 4186 Joules. A BTU is 1055 Joules. Energy is often expressed in watt-hours, or kilowatt-hours, where a watt-hour is 3600 Joules and a kilowatt-hour is 3.6×10^6 Joules. So a kilowatt-hour is 860 kilocalories or 3412 BTU. That means a kilowatt-hour of electricity can heat 100 kilograms of water 8.6°C, or 100 pounds of water 34°F. This is not easy to remember.

Power (energy per second) is also commonly expressed in a variety of units. Watts are simply Joules per second, and kilowatts and megawatts follow naturally. But horsepower? Nowadays a horsepower is defined as 746 watts, or 0.746 kilowatts, which is also hard to remember.

There are very few other units and measures worth memorizing. The international unit of temperature is the degree Celsius (or, less correctly, centigrade), written °C. In the United States, we still use the degree Fahrenheit, °F, where water boils at 212°F or 100°C and freezes at 32°F or 0°C. In practice, water doesn't always boil at 100°C; this is only true when the atmospheric pressure is 1013 milliBar. At higher elevations, water boils at a lower temperature (at 83°C in Denver, for example), but this doesn't detract from the general utility of the Celsius scale. One can easily remember a few fixed points on both scales: 20°C is 68°F, a typical winter indoor temperature; 37°C is 98.6°F or normal body temperature, and -40°C is also -40°F. Scientists also use the Kelvin, or absolute scale of temperature, where the freezing point of water is 273.15°K and the boiling point is 373.15°K. To convert Celsius to Kelvin one simply adds 273.15 (but just adding 273 is close enough).

Finally, there are scientific units associated with atoms and molecules. A mole contains 6.023×10^{23} chemical units (atoms or molecules), and a mole of carbon atoms is defined as 12 grams of carbon. We will go into more detail about this when the need arises.

One metric measure is becoming of increasing interest to weight-conscious Americans, the BMI, or body mass index, which is defined as a person's mass (in kilograms) divided by height, squared, in square meters. Consider a person whose weight is 170 pounds and whose height is 5' 10". In kilograms, the weight (mass, actually!) is just over 77 kilograms. The height is 70 inches, or 177.8 centimeters, or 1.778 meters. Squared, this is 3.16 m^2. Thus, the calculated BMI is $77/3.16 = 24.4$, just under the value (25) that health professionals consider overweight. This sort of calculation could have been done entirely in the U.S. system, but it would still be necessary to convert inches to feet or feet to inches before squaring. A BMI of 25 kg/m^2 corresponds to 5.1 pounds per square foot or 0.0355 pounds per square inch.

One of the biggest advantages of the metric system is that it enables quick mental conversion of unfamiliar units to ones closer to home experience. When I see on television that 100,000 acres of forest are burning in California, my scientific instinct asks, "just how big is 100,000 acres." Divide by 640 (using a calculator, of course!) and I get 156 square miles, or a region 12.5 miles by 12.5. But if the report had said 40,000 hectares, I would know at once, without a calculator, that the area is 400 square kilometers, or a region 20 kilometers by 20. The metric system is faster and easier to use and does not generally require memorization of obscure conversion factors such as 640 acres per square mile. The metric system rewards the inquisitive mind without placing great demands upon it. But I still think it's a pity the French revolutionaries who invented it did not have their way in decimalizing our units of time!

Here are two experiments that illustrate some of the strengths of the metric system, and help keep in mind the critical distinction between energy and power. First, consider the energy you expend climbing a flight of stairs. In ten seconds you gain three meters of altitude. If your mass is 75 kilograms, your change in potential energy is 9.8 x 3 x 75

Joules, or 2200 Joules, and your rate of energy expenditure is 220 watts. That's 0.3 horsepower, which is not a rate of energy expenditure a human being can sustain for very long. Since a Calorie is 4186 Joules, your use of stored food energy would appear to be little more than half a Calorie, but that's assuming 100% efficiency. If your efficiency is more like 33%, you are probably consuming a Calorie and a half – not much. On a diet of 2500 Calories a day, your general rate of energy consumption is about 0.03 Calories a second, or 0.3 Calories in ten seconds. In metric units, that's about ten million Joules per day or 120 Joules per second – 120 watts, that is. So climbing stairs, if you keep at it, will certainly burn off some fat, even though a single flight won't do much good.

The second experiment involves automobiles, and the measurements are a little more precise. At 70 or 75 miles an hour, my car gets 30 miles to the gallon of gasoline, so the car uses about 2.4 gallons per hour. A gallon of gasoline weighs 6 pounds or 2.7 kilograms. The energy content of gasoline is easily measured by burning a small sample and measuring the release of heat; a gallon of gasoline has a heat content of about 1.07×10^8 Joules. At 2.4 gallons an hour, that's 71×10^3 Joules per second, or 71,000 watts. To find out how much energy the car actually needs to keep moving at 70, I did the following experiment. First, I accelerated to just over 75, and then put the car in neutral. When the car slowed down to 75, my partner in this experiment started a stopwatch; when my speed reached 70, she checked the time.

We repeated this experiment a dozen times, trying to make sure that the road was quite flat (and empty!) each time we ran the test. We found, on average, that it took 6.5 seconds for the car to decelerate from 75 to 70.

With two passengers and a full tank of gas, the mass of my car is about 1500 kilograms. At 75 miles an hour, its speed is 33.53 meters per second; at 70, that's 31.29 meters per second. At the beginning of the test, the kinetic energy of the car is 0.843×10^6 Joules; at the end, 0.734×10^6 Joules. So the energy loss is 0.109×10^6 Joules in 6.5 seconds, or 16,800 Joules per second (watts). That's the power required to keep the car moving at an average speed of 72.5 mph. In conventional automotive terms, that's 22.5 horsepower. As for efficiency, that's simply 16,800/71,000, or 24%. Of course, we did the experiment with the air conditioning off so that most of the energy from the engine was actually being used to power the car. By doing all the calculations in a consistent set of units, it was easy to obtain meaningful results.

Proper use of metric units is also helpful in another, more didactic way: If you use metric units consistently, it is harder to make careless mental errors, dividing when you should be multiplying, for example. From the kinetic energy formula $E = \frac{1}{2} mv^2$, we know that energy has units of mass (kg) times distance squared (m^2) divided by time squared (s^2), where v is velocity, m/s. On the other hand, power is energy per unit

13

time, i.e., $(kg)/(m^2\,s^3)$. A watt is a unit of power; a watt-second, or a Joule, a unit of energy. "Weight" is mass times the acceleration of gravity, and since acceleration has units m/s^2, weight has units $(kg)(m)/s^2$. Pressure is weight per unit area, so the units are $(kg)(m)/(m^2 s^2)$ or simply kg/ms^2. Force is mass times acceleration, so the units are $(kg)(m)/s^2$, which is the same as weight; pressure is, equivalently, force per unit area. It's fine to use the terms "weight" and "mass" interchangeably in most contexts, but be careful! Keeping this sort of thing in mind is not hard, and helps make clear the distinctions needed to understand new and evolving technologies. A car needs power (watts; horsepower; energy per second) to accelerate quickly or climb a hill; it needs stored energy (Joules; BTUs, gasoline) to travel a long distance. A community needs a generator to deliver power to homes and businesses; power is reckoned in kilowatts or megawatts (a standard assumption is a kilowatt per household). A household needs power to run the lights and air conditioners and refrigerators but is billed according to the energy used (power times time, or kilowatt hours). These distinctions are routinely screwed up in newspaper reports, but as long as you understand the units involved you can question the reports and, usually, unscramble the mistakes. Or switch to better sources of information. Science begins with precise and carefully defined terms.

Summary: Here is a list of the Top Ten Things you need to have learned from this chapter. They will all be used elsewhere in this book. (But if you are not a U.S. resident you can gleefully skip the last two!)

1. The circumference of the earth is 40,000 kilometers or 40,000,000 (40 million) meters. This isn't so much a property of the earth as it is a definition of the meter.
2. The density of water is 1000 kilograms per cubic meter (or one gram per cubic centimeter). This isn't so much a property of water as it is a definition of the kilogram. The mass of a liter of water is a kilogram, which defines the liter.
3. In the metric system, area is square meters (m^2), and volume is cubic meters (m^3). Velocity (speed) is meters per second (m/s), and acceleration is meters per second per second (or meters per second squared or meters per square second), (m/s^2).
4. The gravity of the earth exerts a force on all objects. This force, also called "weight," is found by multiplying the mass of an object (in kilograms) by the acceleration of gravity, which is 9.81 meters per second per second (9.81 m/s^2) at the surface of the earth. Force is measured in Newtons; one Newton is a kilogram-meter per square second, i.e., $(kg)m/s^2$.
5. Pressure is force per unit area. There are two common units of pressure in the metric system, the Pascal (which is one Newton per square meter, so its units are $(kg)/(ms^2)$, and the Bar (which is 10^5 Pascal). Normal atmospheric pressure at sea level is 1.013 Bar, i.e., very nearly one Bar. Atmospheric pressure is also measured in milliBars, so that normal sea level pressure is 1013 milliBar.

6. There are many forms of energy; all forms can be converted to heat energy. The most useful forms are kinetic energy, gravitational potential energy, and electrical energy. The kinetic energy in a moving object is given by the formula $E = \frac{1}{2} mv^2$, where m is mass in kilograms and v is velocity in meters per second. (We really should write $E = \frac{1}{2} (kg)v^2$, reserving the symbol m for length in meters. But m is also sometimes used to mean mass. Sorry.) The metric unit of energy is the Joule, and its units are $(kg)m^2/s^2$. Gravitational potential energy is given by the formula (kg) **g** h or (kg)**gh**, where **g** is the acceleration of gravity and h is height in meters. The units, of course, are $(kg)m^2/s^2$. Electrical energy is usually measured in watt-hours or kilowatt-hours, but the conversion is easy because a Joule is a watt-second. Or a watt is a Joule per second, which is another way of saying the same thing. A watt is also an electrical current of one ampere at a voltage of one volt. The watt is a measure of power, not energy; power is energy per second.

7. The degree Celsius is the metric unit of temperature (°C). The freezing point of water is 0 °C; the boiling point is 100 °C at normal sea level atmospheric pressure. Again, these are not so much properties of water as they are definitions of the Celsius scale. Scientists also use the Kelvin (or absolute) temperature scale, which is obtained by adding 273 to the Celsius scale, so that, for example, water boils at 373 K.

8. In the metric system, thermal energy is also measured in Joules, but some older units are still in common use. The British Thermal Unit (BTU) is the energy needed to raise the temperature of a pound of water by one degree Fahrenheit; the calorie is the energy

needed to raise the temperature of a gram of water by one degree Celsius, and the kilocalorie (or food Calorie) is the energy needed to raise the temperature of a kilogram of water by one degree Celsius. A Calorie (that is, a kilocalorie) is 4186 Joules. A BTU is 1055 Joules.

9. In the United States (and nowhere else) we use inches, feet, pounds and other obsolete units. If you can remember that there are twelve inches in a foot, three feet in a yard and 5,280 feet in a mile, then surely you can also remember that an inch is legally defined as 2.54 centimeters (or 0.0254 meters), from which everything else can be calculated if you must. A meter is $100/2.54 = 39.37$ inches, more or less, just over a yard. And a kilometer is about six-tenths of a mile. For liquid measure, a quart is about 0.95 liters.

10. To convert degrees Fahrenheit to Celsius, subtract 32, multiply by five, and divide by nine. Ugh! To convert Celsius to Fahrenheit, multiply by nine, divide by five, and add 32.

Problems and Exercises

Exercise #1: Using only the information given in this chapter, what is the radius of the earth in kilometers?

Exercise #2: If water were used in a barometer instead of mercury, what would be the minimum height of the water column?

Exercise #3: What is the total pressure underwater at a depth of ten meters?

Exercise #4: What power is required to pump one liter of water per second up a hill one hundred meters high?

Exercise #5: Human beings float, barely. If your lungs are full of water, you sink. Roughly speaking, what is the density of your body? What is your volume?

Exercise #6: (This is sort of a trick question!) How many liters of gasoline are in a cubic meter of gasoline?

Exercise #7: The heat of combustion of gasoline is about 48,000 Joules per gram of gasoline. If you could somehow burn a gram of gasoline underwater, in, say, a minuscule stove contained in a one-liter water bottle, by how much would the temperature of the water increase?

Exercise #8: The heat of combustion of sugar is about 17,000 Joules per gram of sugar. What would the results be in the previous exercise if you burned a gram of sugar instead of a gram of gasoline?

Part I: Politics and Tides: Things that Change with Time

Chapter 1: Sun, Moon, Earth, and Tides

During the summer my wife and I spend most of our time on an island off the coast of Maine. On the dining room wall of our cottage is a tide clock, designed to look much like an ordinary clock except that it has only one hand and no numbers. Where the number twelve should be it says "High," and in the six position it says "Low." At three o'clock it says "Falling," and at nine it says "Rising." The hand goes round the clock once every 12 hours and 50 minutes[1].

At ten in the morning, I glance at the clock. The hand points where the number two should be: It's about two hours past high tide, and the tide is going out. In roughly four more hours, at two in the afternoon, it will be low tide. That would be a good time to collect mussels or dig for clams, weather permitting. The tide clock doesn't say how low the low tide will be -- it's a cheap and unsophisticated machine -- so I consult the local tide tables online. If the height of the low tide is way below normal, we'll have mussels for dinner. It's so much easier to collect mussels if the tide is way out. And our feet won't get so cold and wet. But if it's not such a low

---

[1] The time between successive high tides is not exactly 12 hours and fifty minutes every day: This is the average. The exact interval on any one day can even be off by an hour. But the tide drops only a little in the half hour before and the half hour after the peak.

tide, or if we get a late start, we'll dig for clams instead. Clams inhabit the mid-tide beach and can be harvested any time between the falling and rising half-tides. Mussels prefer to live at the low tide mark.

We also check the clock if we're going ashore to get groceries and supplies. High tide, on average, is ten feet above low; lugging stuff up a steep ramp can be avoided by planning trips around the tides. We watch the tide when we're sailing, too: Every sailor does. The safest time to be out in a boat is just after low tide: If you run aground, the tide will quickly lift you off. The most dangerous time is when the tide is high and falling: Grounding out will leave you high and dry, and pounding waves can destroy a boat long before the next high tide. The higher the tide when you run aground, the more precarious your situation, and the ultimate accident is grounding out on a spring tide[2], when the moon is full (or new) and the tides are the highest of the month. I've seen it done. The captain had two options: wait a month for the next spring tide, or get a quick and violent tow while the tide was just beginning to fall. He wisely chose to accept a tow and sacrifice part of his rudder and keel.

Tidal currents are important, too, especially if your boat is small and slow. Roughly halfway between high tide and low,

_____

[2] Spring tides have nothing to do with the four seasons. There is a spring tide every month! The term merely means the highest tide of the month, although some dictionaries say there are two spring tides each month.

the currents reach a maximum, and they can either speed you on your way or stop you altogether. These currents, like the tidal heights, vary from place to place around the world; where we live, they often reach one or two knots. The hull speed of our boat is barely five.

Some years ago I was in Seattle on business. Two of my clients were avid sailors, and soon we were trading stories. But something they said puzzled me. They spoke of four sorts of tides, a high high tide, a low low tide, a high low tide, and a low high tide. So I asked for an explanation. The answer I received went something like this: "Well, there's the sun and the moon, and the moon has the biggest influence on the tides, but the sun also has an effect, and that's why the two high tides have different heights." That's fine, I said, but in Maine we also have the sun and the moon, and the two high tides are just about equal.

There is a fairly simple (but not entirely correct!) explanation for the Seattle tides. The city lies at the eastern end of the San Juan de Fuca Strait, and the tides of the Pacific Ocean have to pass through the strait to reach the city. The strait acts like a big bathtub. Water sloshes back and forth, from one end to the other. Like a bathtub, there is an inherent natural frequency to this sloshing, determined by the average depth of the water. In a bathtub, children can make the water slosh out of the tub by moving back and forth at just the right intervals. Not any motion will do; the trick is to be in sync

23

with the water. Physicists call this effect "resonance." The earth's rotation under the sun and moon provide a force that moves the water of the oceans, but this force is not in sync with the natural frequency of the San Juan de Fuca Strait.

The speed of a bathtub or ocean wave is given by the formula $c = \sqrt{(gh)}$, where h is the depth of the water and $\mathbf{g}$ is the acceleration of gravity (32 feet per second per second or 9.8 meters per second per second). If the depth of the bathtub is half a foot, the speed is 4 feet per second.

This explanation satisfied me for a long time. But I still didn't understand why there are two high tides every day, not one. One high tide occurs roughly when the moon is at its zenith; the other when the moon is on the far side of the earth. When the moon is full, the moon is at its zenith around midnight, and, sure enough, there is (in Maine!) a high tide near midnight, but why is there also a high tide around noon?

The standard explanation is both simple and subtle. Imagine that the earth has no continents at all, that the planet is "flooded" (think of Noah). Water that is on the side of the planet nearest the moon is pulled upward toward the moon. That much is clear. Because the water is nearer to the moon than the average place on earth -- that is, nearer than the center of the earth -- the moon has a stronger tug on this part of the ocean than its tug on the center of the earth.

Effectively, the earth's gravity on this part of the ocean is weakened by the tug of the moon. On the opposite side of the earth (see Figure 1.1), the moon has a weaker than average tug on the ocean, because the water is farther away from the moon than the center of the earth. Once again, the earth's gravity is effectively weakened because the tug of the moon is weaker than at the center of the earth. Because the sum of the attractive forces of the earth and the moon is effectively weaker (compared to the poles or the center of the earth) at both places, the tide is high at both. And, halfway in between, at the poles, where the moon's tug is the same as at the center of the earth, the tides are low, giving Noah's planet two high tides and two low tides daily[3].

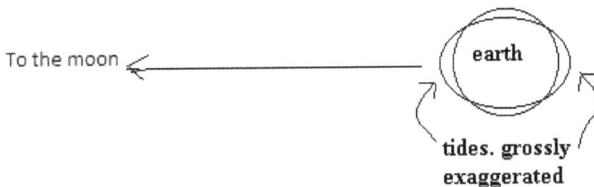

To the moon   earth

tides. grossly exaggerated

Figure 1.1 Tides: The Moon and the Oceans

This explanation can be fleshed out in detail. It turns out that the effect of the sun is about half (46%) that of the moon, which explains why there are spring tides twice a month, around the time of the new moon and the full moon, when the sun and the moon line up and their effects add up. Furthermore, the moon's orbit is eccentric ($\varepsilon = 0.055$), which means the moon is 5.5% closer to the earth than average at

---

[3] The mathematical details of this argument are discussed in Exercise 4.

perigee, and 5.5% farther at apogee. This can change the gravitational effect of the moon by roughly 20%, both positively at perigee and negatively at apogee. Likewise, the earth's orbit around the sun is also eccentric ($\varepsilon = 0.017$), but the effect on the tides is much smaller.

Unfortunately for the theorists, but fortunately for the rest of us, the waters of the earth receded, Noah's ark went aground, and the earth is not a flooded planet. So the theorists have an excuse for why some of their predictions are wrong. For example, the theory predicts that the highest possible tide (spring tide at perigee) will be about three feet. It also predicts the highest tides will be along the equator, although in fact, equatorial tides are notoriously small. Furthermore, some places on our planet have only one high tide a day, and in some places the spring high tides do not occur at noon and midnight. Whoops!

There are other difficulties, too. The earth's axis is not perpendicular to its motion around the sun ("the ecliptic") but is tilted at an angle of 23.5°. The moon's orbit is also tilted, at an angle of 5° with respect to the ecliptic. And there are continents in the oceans. So, while it is fairly easy to understand the major forces that drive the tides -- the moon, the sun, the rotation of the earth with respect to the moon and sun, and the changing distance between the earth and moon -- the effects of these forces are quite complex.

The flooded planet theorists have an even bigger problem. The average depth of the oceans is about 4000 meters. Using the boxed formula given on page 24, the maximum speed of an ocean wave is about 200 meters per second (speed equals the square root of 9.8 times 4000). At the equator, the earth rotates at a speed of 463 meters per second. So it is impossible for a wave to move fast enough to keep up with the earth's rotation.

Here are some useful ways of thinking about the size of the earth. As noted earlier, the meter was originally defined so that the distance between the North Pole and the equator is ten million meters, for a circumference of forty million meters, or forty thousand kilometers. (The earth is not quite a sphere, so this didn't quite work out.) Equivalently, the nautical mile was defined to be one minute of arc, i.e., one-sixtieth of a degree or sixty nautical miles per degree, or 60 x 360 = 21,600 nautical miles circumference, which also isn't quite exact. It is also useful to know that a nautical mile is about 6000 yards or 1.15 U.S. miles.

Along the northern Atlantic coast, the ocean sloshes back and forth between Europe and America, more or less confined in a giant deep bathtub that is about a sixth of the circumference of the earth in length. There is a point about halfway across the Atlantic, known as the amphidromic point, that has no tides at all. The waters rock back and forth on this point like a rotating seesaw. The natural frequency of this slosh is about the same as the (effective) twelve and a half-hour cycle of the earth's rotation under the moon, and this is probably why the tides in northern New England are so high. But in truth the height of the tides everywhere on earth depends on the size and shape and nature of the "local" bathtub, and, as we have seen in the case of Seattle, the bathtub can have peculiar effects. It can cause the tides to lag so that high tides occur well after noon and midnight, and it can cause the tides to be so uneven that only one high tide per day is seen. This is not to say that tides are unpredictable: they are the most predictable form of weather on the planet. But prediction is generally quite complicated, involving many different factors, and the apparent simplicity of the New England tides is accidental.

Problems and Exercises

Exercise #1: How deep would the oceans of Noah's planet have to be for the speed of the tides to match the earth's rotation with respect to the moon? Halfway around the earth is twenty million meters or 10,800 nautical miles. To travel this distance in twelve hours and twenty-five minutes requires a speed of 433 meters per second or 1400 feet per second. Using the formula speed = $\sqrt{gh}$, where g is the acceleration of gravity (32 feet per second per second or 9.8 meters per second per second), the depth of the ocean (h) would have to be over 60,000 feet (19,000 meters). In fact, the average depth of the oceans is about 13,000 feet or 4,000 meters.

Further Information: For more information on tidal theory, use the web and search for amphidromic point.

Exercise #2 The weight of the earth: The gravitational force between two masses is given by Newton's formula $F = Gm_1m_2/r^2$, where m_1 and m_2 are the masses, r is the distance between them, and G is the universal gravitational constant. Since force is also equal to mass times acceleration ($F = ma$), the acceleration experienced by a small mass m attracted by a much larger mass M is $a = GM/r^2$. For objects on the surface of the earth, a is easily measured by dropping a stone off the top of the Leaning Tower of Pisa; its value, as noted above, is 9.8 meters per second per second. If we knew the value of G, we could calculate the weight of the earth! In 1798, Henry Cavendish did the first direct measurement of G in a very delicate experiment that directly measured the (very small) force between two heavy spheres of known mass. The result is $G = 6.6 \times 10^{-11}$ (meters)^3 per kilogram per (second)^2. Verify that the mass of the earth is 6.0×10^{24} kilograms.

Once the mass of the earth is known, the masses of the sun and the moon can be calculated if the distances are known. The mass of the sun is $M_S = 3.3 \times 10^5 M_E$, and the mass of the moon is $M_M = 0.0123 M_E$, where M_E is the mass of the earth.

Exercise #3: Changes in local gravity: The force experienced by a small object on the surface of the earth decreases with altitude. We can calculate this change by differentiating the equation $a = GM/r^2$ with respect to r. The result is $\Delta a = -2GM/r^3 \, \Delta r$, where Δa is the change in local acceleration **g** and Δr is the change in distance from the center of the earth. At 500 meters above the surface of the earth, verify that **g** decreases from 9.8 meters per second per second by about 1.5×10^{-3} meters per second per second, that is, by about 0.015%.

If you weighed yourself (on a spring scale!) at the top of a five hundred meter tall building, by how much would your apparent weight change? What if you used a balance-beam scale, like the one found in most doctors' offices?

Exercise #4: Changes in apparent local gravity caused by the sun and moon: As noted in this chapter, the effect of gravity at the surface of the earth is weakened when the sun or moon is directly overhead (compared to the effect when the sun or moon is on the horizon). Use the formula $\Delta a = -2GM/r^3 \, \Delta r$ to calculate the apparent change in local acceleration caused by the sun and the moon. (Here Δr is the radius of the earth, while r is the earth-sun (1.5×10^{11} meters) or earth-moon (3.8×10^8 meters) distance. How do these changes compare to the effect of a simple change in altitude? By how much does your apparent weight (on a spring scale!) decrease when the moon is overhead compared to when the moon is on the horizon?

Exercise #5: Measuring g: If a baseball is dropped from a height of four feet, it takes exactly half a second to hit the ground. (a) Show that this experiment measures **g**. (b) A major-league fastball takes half a second to reach home plate. Does it fall four feet?

Exercise #6: Orbits. Based largely on the work of Kepler, Isaac Newton showed that the mass of a large object like the sun can be determined by observing the motion of a satellite (or planet) in orbit around it: $M = 4\pi^2 R^3/(GT^2)$, where R is the distance of the planet from the sun, T is the orbital period, and G is the universal gravitational constant. Orbital periods are easy to measure; planetary distances are not. How was the distance of the earth from the sun first measured?

Exercise #7: Geostationary Satellites. A geostationary satellite orbits around the earth at the same rate that the earth spins, that is, its period is 24 hours. Use the information provided in this chapter to calculate the distance of such a satellite from (a) the center of the earth and (b) from the surface of the earth. NOTE: The period of a satellite is given by the formula $T = 2\pi\sqrt{(R^3/MG)}$, where R is the distance of the satellite from the center of the earth, M is the mass of the earth, and G is the universal gravitational constant. It is not really necessary to know M and G separately, since, at the surface of the earth, $\mathbf{g} = MG/R^2 = 9.8 \ m/s^2$, where (in this case) R is the radius of the earth.

Exercise #8: Low-Earth-Orbit (LEO) Satellites. A typical LEO satellite orbits around the earth in 90 minutes. Using the same methods as in the previous exercise, calculate the satellite's altitude.

Exercise #9: Communication Delays. While watching the Red Sox play the World Series in Fenway Park, I turned down the sound on the (cable) television and listened to the play-by-play on the local Boston radio station. Sometimes this was a big mistake: The radio announced the plays before they showed up on the TV! I assume this is because the TV signals were sometimes (but not always) transmitted via a geostationary satellite. The speed of light is 3.0×10^8 m/s. Using the results of exercise #7, calculate how long it takes a signal to make the round trip to such a satellite.

Some Solutions

Exercise #4: The distance between the earth and the moon is 3.8×10^8 meters, while the mass of the moon is $0.0123 \times 6.0 \times 10^{24}$ kg. ($M_M = 0.0123\ M_E$.) Thus, from the formula given in the exercise, $\Delta a = -2 \times 0.0123 \times 6.0 \times 10^{24} \times 6.6 \times 10^{-11} \times 6366 \times 10^3/(3.8 \times 10^8)^3 = -1.13 \times 10^{-6}$ m/s^2. This is very small compared to **g**, 9.81 m/s^2. However, the decrease in **g** caused by raising an object one meter, 0.3×10^{-6} m/s^2, is entirely comparable, which is why the moon causes the oceans to gain altitude through tidal motion.

Note that if this formula is applied to oceans on the far side of the earth (the right-hand side in Figure1.1), the result is the same: The effect of the moon is to diminish **g** so that once again the tide is high.

As for the sun, we replace $M_M = 0.123$ M_E with $M_S = 3.35$ x 10^5 M_E, and replace the earth-moon distance, 3.8 x 10^8 meters, with the earth-sun distance, 1.5 x 10^{11} meters. The result is $\Delta a = -2$ x 3.3 x 10^5 x 6.0 x 10^{24} x 6.6 x 10^{-11} x 6366 x $10^3/(1.5$ x $10^{11})^3 = -0.49$ x 10^{-6} m/s^2. So the maximum effect of the sun is a little less than half that of the moon.

While your change in weight at the top of a 500-meter building amounts to about 0.015% (about 10 grams), the effect of the moon is only about 7 milligrams.

Exercise #5: The velocity of a falling object at the surface of the earth is given by the formula v = **g**t, where t is the time passed since the object was dropped. The distance the object travels is given by the formula distance = ½**g**t^2. Thus, knowing that a dropped stone falls four feet in half a second, we can calculate that **g** = 32 feet per second per second.

But a major-league fastball is a whole different kettle of fish. When an object is moving at high speeds, air resistance must be taken into account. A glider (or airplane with engines shut down) does not immediately fall from the sky but can maintain altitude (or even climb!) simply by slowing down. A spinning fastball can do much the same. It can climb, it can dive, or curve inside or out. Only in a vacuum would a fastball drop four feet, and if the pitcher threw it with a sidearm or underarm motion so that the initial velocity was slightly upwards, it wouldn't have to fall at all in the half-second before it crosses the plate.

Chapter 2: Thinking about the Weather

Here's a trick question: What is the third most common gas in the atmosphere? Everyone knows that nitrogen (N_2) is the largest component, at 78%, followed by oxygen (O_2) at 21%. Most people who do not have a background in science guess carbon dioxide is next, but CO_2 is only 0.04% of the air. Those who know some science guess argon (A), which is half right: Argon makes up a little under one percent of dry air. But the atmosphere isn't dry: On average, water vapor (H_2O) makes about one percent, but the proportion is extremely variable.

The daily weather reports provide the water content of air in a disguised sort of way, when they note the dew point, which is the temperature at which water would start to condense (and form rain or fog) if the air were cooled. Let's say the dew point is 68°F, or 20°C -- a muggy day. According to the Handbook of Chemistry and Physics, the vapor pressure of water at 20°C is 17.5 millimeters of mercury, compared to normal atmospheric pressure at sea level, which is 760 millimeters of mercury. So, under these conditions, the atmosphere is 17.5/760 x 100 or 2.3% water vapor. At a dewpoint of 10°C (50°F), the water vapor pressure is only 9.2 millimeters, but that's still 1.2% of the air[4].

The curve describing dew point versus pressure is shown in Figure 2.1. Roughly speaking, the logarithm of the pressure (log (P)) is proportional to the inverse of the absolute temperature T, where T is the temperature in degrees Celsius plus 273.15. This is true for any liquid that can form a gas, although the slopes vary depending on the liquid.

Fig. 2.1: Vapor Pressure of Water

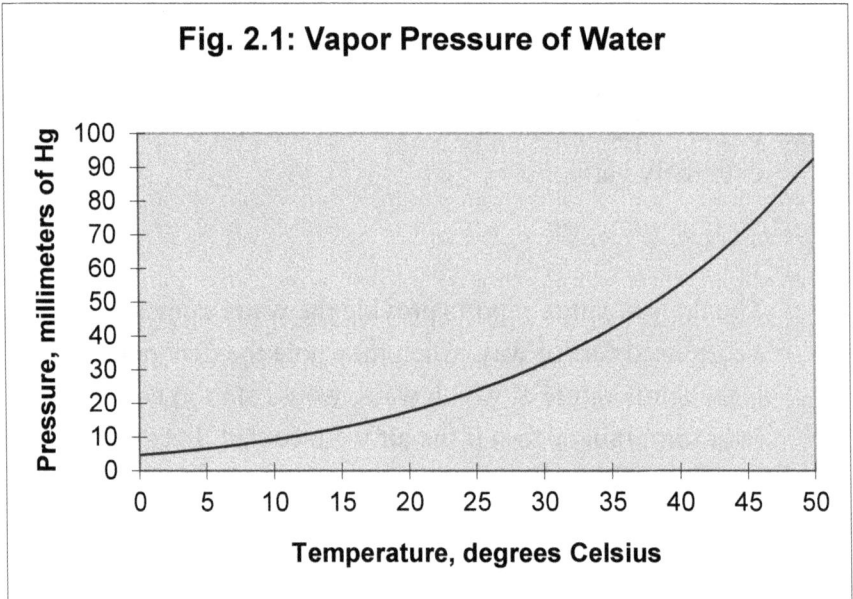

---

[4]According to the October 2002 issue of AOPA Pilot, the highest dew point ever recorded in the United States was 90°F, or 36 mm -- 5% water.

Fig. 2.2: Water Vapor Pressure

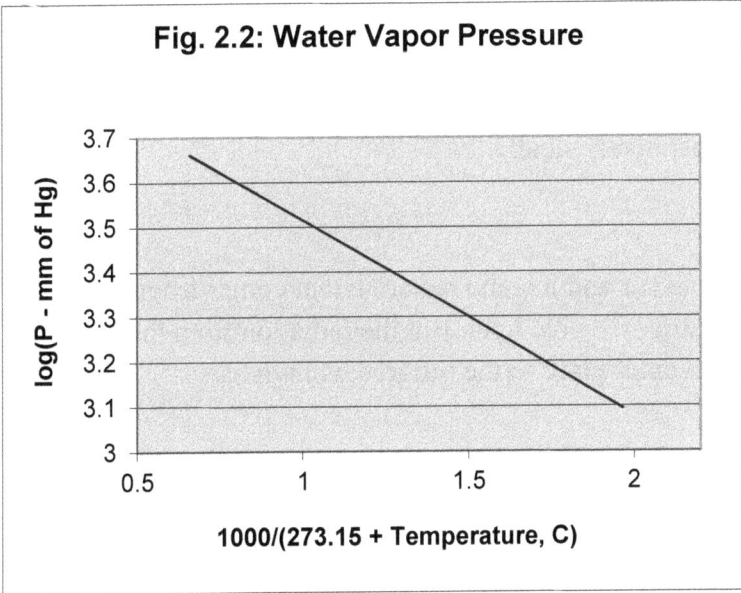

Obviously, warm air generally has more water content than cold air, so the atmosphere contains more water in summer than in winter. Likewise, the lower part of the atmosphere is wetter than the upper part, and indeed the reason it rains is that warm wet air cools when it rises and if it cools enough, to the dew point, the water vapor condenses.

Water vapor is a greenhouse gas, indeed, the most important greenhouse gas. What do we mean by greenhouse gas? The concept is a little tricky. To understand it, we first have to understand that the earth loses heat by radiation. Every bit of energy that the earth receives from the sun is radiated back into space, more or less immediately. (By more or less I

mean within seconds or minutes, or at most within a year!)
Every warm object radiates heat energy, but for things not
surrounded by a vacuum, radiation losses are usually
comparatively small.

We can see and feel the radiation that comes from the sun in
the form of visible light. But the radiation from the earth,
which takes place in the infrared, is invisible.

The earth's atmosphere is transparent to the light of the sun:
Nitrogen, oxygen, water, argon, and carbon dioxide do not
absorb visible sunlight. (Ozone, O_3, which forms at the top
of the atmosphere, absorbs ultraviolet light, providing some
protection from skin cancer.) Nitrogen, oxygen, and argon
are also transparent to the invisible infrared radiation that the
earth emits back into space. But not water and carbon
dioxide, both of which are strong absorbers of infrared.
There isn't much CO_2 in the air, so most of the infrared gets
through. With water vapor, it's a different story.

Here's how the greenhouse effect works. A water molecule
in the air absorbs a photon of infrared light emitted by the
earth. This heats up the water molecule for a while, and then
the molecule radiates another photon and cools back down.
The emitted photon can be sent off in any direction: Out to
space, back to the earth, or off to the side. The photon may

escape altogether or be absorbed by another water molecule. The net result is that the air warms up, and heat losses from the earth are reduced as photons from the water molecules are radiated back down on the planet.

This effect is often compared to the effects of a blanket on a bed, or for that matter to the glass in a greenhouse. These are not very good analogies, because neither involves radiation. Never mind. Both analogies correctly say that heat is trapped, and trapped heat causes a rise in temperature.

Now you know why dry nights are chilly and humid nights are less so. But the effects of water are far more elaborate and complicated. Snow, ice, rain, and clouds are all made of water, and they all affect the climate. Snow and ice are the easiest to consider, although their effects are quite strange. Obviously, snow and ice form when it is cold. Cold air has less water vapor in it, so the greenhouse effect is reduced -- and the air has a tendency to get even colder. Likewise, snow and ice reflect sunlight, so, during the day, the cold earth radiates visible light out to space, and gets even colder. Cold produces more cold when the earth is snow-covered. There is even some evidence in the geological record that the earth was once a giant snowball. Cold enhances cooling. This is called a positive feedback effect.

Just the opposite is true in the tropics, where there is no snow. The hotter the air, the more water vapor it contains; the more water vapor, the greater the greenhouse effect, and the hotter the air. Heat enhances heating, another positive feedback.

Clouds, fortunately, are even stranger. The tops of clouds reflect sunlight, just like snow. So clouds produce cooling. But clouds also trap heat radiated from the earth. So clouds produce heating.

Are you confused? You should be. Weather is complicated. Scientists are used to dealing with systems that have negative feedback, meaning that the things we like to study have a sort of frictional component, acting in the opposite direction of the main force. But the climate has two major components with positive feedback that act in the same direction as the main force. Fortunately, these two effects -- "runaway cooling" at the poles and "runaway heating" at the tropics -- oppose each other and produce the current climate. But it seems like a delicate and unstable balance, and there is much evidence that the earth has experienced huge changes in climate over the eons, due no doubt to small changes in the earth's orbit and tilt. And that is why scientists are nervous about the increasing carbon dioxide content of the atmosphere. Small changes can have big effects, and the earth's atmosphere is unstable. But it is also why there is so much disagreement about how big the effects are likely to be, and how quickly they are likely to occur. The balance of effects is so delicate and tricky that accurate prediction is extremely hard.

One might expect that the effects of carbon dioxide should be greatest when there is relatively little water vapor in the air, i.e., in places and times when it is cold. There is some evidence that this is indeed true: Measurements of global warming to date seem to indicate that the greatest warming occurs at the poles and, in the rest of the world, at night. Since daily high temperatures occur in the afternoon, while daily lows occur at night, one might expect global warming to proceed by an increase in the average lows rather than the average highs. This does indeed seem to be the case[5].

Water vapor is also enormously important to the daily climate: It rains! To understand just why takes a little work. During a clear day, when the sun shines, the hottest part of the atmosphere is near the ground, where the sun's rays are absorbed. At night, just the opposite should be true: The air is (or should be!) coldest nearest the ground. This implies that there is an inversion point somewhere high in the atmosphere, where the temperature remains more or less constant. Nice theory, but it isn't usually true in practice. The

---

[5] Freeman Dyson, New York Review of Books, Vol. L, #8, p. 4, 5/15/03.

earth radiates a prodigious amount of heat into space, and even on a clear day there is usually enough water vapor in the air to insure that temperature decreases with altitude.

There are four ways the surface of the earth can lose heat, conduction (this is the heat loss you feel when you sit on a cold bench), convection (the heat loss you feel in a cold wind), evaporation of water, and radiation (the heat gain you feel on a sunny day, especially a cold sunny day). It turns out that conduction is largely ineffective for gases -- air is a pretty good insulator if it isn't moving -- so convection and evaporation and radiation are the main methods for transferring heat in the atmosphere. As a rough rule, the temperature decreases about 6.6° C for every 1,000 meters of elevation, or 3.5° F per thousand feet, and temperature inversions are rare except on the downwind side of mountains, for reasons we will soon explore.

HEADNOTE:[6] The next couple of pages contain the most mathematically complex material in this book. Understanding the equations that follow requires a firm background in calculus. If the reader is not comfortable with calculus, he or she should simply skip (or, better, skim) the equations. All they say, in the end, is that hot air rises, except when it doesn't. That is, they show under what atmospheric conditions heated air will or will not rise. For those familiar with calculus, these equations are absolutely elegant, an aesthetic delight. The reader who skips or skims will miss the beauty of the analysis, but not the important implications of the result.

One equation that should not be skimmed is the Ideal Gas Law, $PV = nRT$. This formula is as important to understanding gases as is Newton's Law of gravitation to understanding planets and satellites. The equation interrelates the pressure P, (absolute) temperature T, and volume V of n moles of a gas. Unlike Newton's Law, the ideal gas equation is not exact. It is an excellent approximation for real gases under most circumstances, which is a long way of saying that it is valid except when it isn't. It is least likely to be accurate when temperatures are low, pressures are high, and when the gas contains molecules that attract or repel each other strongly. Conversely, it works best at low pressures, at high temperatures, well above the boiling point, and for molecules that interact only weakly. Helium at room temperature is as ideal a gas as there is. But here's a surprise: The interior of a star, which consists mostly of highly interacting electrons and protons – a plasma – is well-described by this equation. Stars are so hot that the equation works well!

---

[6] A "headnote" is an interruption in the text that is too important to leave to a footnote!

Air pressure decreases with altitude, just as water pressure increases with depth. The pressure at any elevation depends on the total weight of the air above. A little calculus goes a long way here:

$$dP/dz = -\rho\mathbf{g} \qquad (1)$$

where P is pressure, ρ is air density, z is height, and $\mathbf{g}$ is the acceleration due to gravity. For a liquid, ρ is essentially constant with depth, but for air (and for almost any gas), the ideal gas equation applies, i.e.,

$$PV = nRT \qquad (2)$$

where V is volume, T is the absolute temperature, R is a constant ("the gas constant"), and n is the number of moles of gas in volume V. The ideal gas equation can be re-written as

$$n/V = P/(RT) \qquad \text{or} \qquad \rho = Pm/(RT)$$

where m is the molecular weight of the gas (roughly 29 grams per mole for air, that is, 0.029 kilograms per mole). Substituting,

$$dP/dz = \mathbf{g}Pm/(RT)$$

This can be integrated to give

$$\log_e (P) = A \ \text{-}\mathbf{g}zm/(RT)$$

where A is a constant such that $\log_e (P_O) = A$ at sea level where z is zero. Put another way,

$$\log_e(P/P_O) = \text{-}\mathbf{g}zm/(RT)$$

where P is the pressure at altitude z and P_O is the pressure at sea level.

With $\mathbf{g}$ equal to 9.8 meters per second per second, T set equal to 300 K (roughly 80° F), and R equal to 8.31 Joules per degree per mole, at 1600 meters above sea level, or one mile (e.g., Denver),

$$\log_e (P/P_0) = -0.18$$

or $P/P_0 = 0.83$

Thus, the atmospheric pressure at Denver is only 83% of the sea level pressure. In the course of putting together these equations we have assumed the temperature does not change with altitude, which is not quite right, but is accurate enough since pressure changes so much more rapidly than temperature, which might drop 10° C from 300 K to 290 K (i.e., from 27°C to 17°C) in a 1600 meter change of altitude.

Everything in the boxed section above is irrelevant to our current purpose, which is to understand something about the weather. The next equation, which arises in the science of thermodynamics, is much more important. For a gas that expands adiabatically, that is, without being heated or cooled by an outside source,

$$dT/dP = C_P/\rho \qquad (3)$$

where ρ is the gas density as before, and C_P is the heat capacity of the gas (in Joules per kilogram degree) at constant pressure. Several remarks are in order here. First, any gas that expands rapidly expands adiabatically[7]: The assumption is that heat transfer from an outside source is slow, while expansion (or compression) can occur much faster. Second, "heat capacity" means the amount of energy a substance absorbs per degree of temperature (when the pressure is held constant). It is a well-known, well-studied and well-understood property of all materials, although its value varies from material to material.

We can now combine equations (1) and (3). The result is

$$dT/dz = g/C_p$$

which is a remarkably simple equation. With g equal to 9.8 meters per second per second and C_P (for air) equal to 1000 Joules per kilogram per degree, dT/dz is constant at about 0.01°C per meter, or 1°C per 100 meters or 10°C per kilometer or 5.4°F per 1000 feet.

---

[7] The word "adiabatic" simply means without an external source of heating or cooling.

We have to be very careful about the meaning of this equation. This is **not** the rate at which the temperature of the atmosphere changes with altitude. And this equation does **not** apply to moist air. What this equation says is that if a "packet" of dry air is elevated, it will cool by just this extent. A rising packet of air **must** cool in exactly this way.

The situation for damp air is almost the same, up to a point. When the air has cooled to the dew point, water vapor begins to condense, and water droplets (fog) begin to form. When any vapor condenses, it releases heat in the process, just as an evaporating liquid takes up heat. As condensation occurs in a packet of rising moist air, the air cools **much** more slowly.

We are now in a position to understand some basic weather patterns. In the morning, before the sun has had a chance to do much heating, the actual decrease in temperature with altitude is relatively small, let's say 5° C for every 1000 meters of altitude. Also, since the morning air is cool, it doesn't contain much water vapor. Imagine a packet of sun-heated air near the ground that begins to rise (hot air, being less dense than cold air, rises). As soon as it rises, it cools at a rate of 10° C per 1000 meters. Whoops! This is colder than the surrounding air, so it drops again, or, more accurately, it never rises in the first place. The atmosphere is stable, there are no rising columns of hot air, and your airplane ride will be very smooth.

Afternoons are different. The actual decrease in temperature with altitude might increase to 15° C per 1000 meters. A packet of air that rises 1000 meters will still cool 10° C, but this is 5° C warmer than the surrounding air, so it will continue to rise. Your airplane ride will be bumpy. Furthermore, since there is nothing to slow down our rising air packet, it will keep climbing until the dew point is reached and fog (clouds) begin to form. This does not slow down the rate of rising; on the contrary, it accelerates it, since our rising packet of air will now cool at a rate much less than 10° C per 1000 meters, and so will be much much warmer than the surrounding air. The result, of course, is towering cumulous clouds, thunderheads, wildly accelerating columns of air, and, when the fog droplets begin to coalesce, rain.

The difference between stable air and unstable air is like a switch. Warmed air either rises or it falls. As you may imagine, this means that weather prediction can be nearly impossible, although, when the atmosphere is predictably stable, the forecasters usually get it right. Sometimes the weather is easily predictable. Sometimes not. And of course there are huge complications due to huge and low-pressure zones, fronts, and the jet stream.

On the lee side of mountain ranges (e.g., Denver), another adiabatic effect occurs. Air flowing from west to east warms (by 10° C per 1000 meters) as it falls, heating the air far above the ground, producing a temperature inversion so that the air aloft is warmer than the air at ground level. The result can be an exceptionally stable air mass, halting the normal afternoon upward flow. The result, of course, is stagnant air, and pollutants in the lower atmosphere are unable to escape. An excess of stability can have nasty consequences.

On the windward (west) side of mountain ranges, pollutants can also be trapped. An air mass may lack the energy to move across the range, and the air becomes trapped in a recirculating pattern.

Adiabatic effects also explain why the windward sides of mountains are wet while the lee sides are dry. Mountains force the west wind to rise, cooling the air quickly, enhancing condensation. On the lee side, the wind heats and condensation cannot occur. When there is rain on the east side of mountains, it is usually because of a "back-door cold front," when the wind flows from east to west and the mountain effects are reversed.

Weather is complicated. The parts that I have discussed above are the parts that are well understood. Because of complexity, detailed and accurate long-range prediction is thought to be impossible. Small changes in temperature and moisture can have large, uncertain effects. Improved computer models will probably improve forecasting accuracy, but progress is likely to be painfully slow.

Problems and Exercises

Exercise #1. If the atmospheric pressure in Denver, Colorado is only 80% of sea-level pressure, what percentage of the air is water vapor if the dew point is 20°C? Why is this percentage higher than at sea level? How is this consistent with the fact that Denver's climate is very dry?

Exercise #2. If a "packet" of heated air is rising at constant volume and diminishing pressure, does the number of air molecules in the packet (i.e., the mass of the packet) change as it rises?

Exercise #3. Infrared radiation from the surface of the earth, and from water and carbon dioxide molecules in the air, takes place during the day as well as the night. Independently of any rising air currents, how does this affect the temperature of air that is not close to the surface?

Exercise #4. How does the formation of clouds affect the "switch" that determines whether air near the ground rises or stays stationary?

Exercise #5. Air pressure can be described in a great variety of units. Until recently, one of the most common was millimeters of mercury (mm Hg), which relates pressure to the height of a column of mercury in an old-fashioned Torricelli barometer. Airplane pilots use a similar measure, inches of mercury. The normal atmospheric pressure at sea level is 760 mm, or 29.92 inches. (Air traffic controllers say "two niner niner two" in order to improve clarity.) Another unit is pounds per square inch or psi: normal pressure is about 14.7 psi. A more modern unit is the bar, or millibar: Normal pressure is 1013 millibars (1.013 bar). In the metric system, pressure is force (mass times acceleration) per unit area (square meters), so the units of pressure are kilograms per meter per second per second, where one bar is 10^5 kilogram per meter per second per second. Calculate the mass of a column of air above an area of one square meter or one square centimeter. What is the mass of the earth's atmosphere?

Exercise #6. Use the equations and information provided in this chapter to calculate the density of air. How does it compare to the density of water, which is 1000 kilograms per cubic meter (or one gram per cubic centimeter)?

Exercise #7. The intensity of the sun on the equator at noon is approximately 1000 watts per square meter. Using a simple geometric argument, show that the average intensity of the sun, averaged over day, night and year, is 1000/4, which is also the average radiation intensity of the earth out to space.

Exercise #8. According to the Stephan-Boltzmann equation, a heated object in the vacuum of interstellar space radiates energy at a rate of $I = AT^4$ watts per square meter, where T is the absolute temperature in Kelvins and $A = 5.67 \times 10^{-8}$. Assume that $T = 280$ K for an average earth temperature. What is the calculated radiation intensity of the earth? How does this answer compare to that obtained in the previous exercise? Explain the difference. What would the earth's average temperature be in the complete absence of greenhouse effects?

An Extension: Since we have been examining the rapid compression of gases, it makes sense to look at another familiar example where adiabatic effects are important. In the design of diesel engines, a mixture of air and fuel is suddenly compressed to a very small volume. Because of the change in volume, the equation given above for adiabatic effects in the atmosphere cannot be used, but a similar equation,

$$dT/dV = -C_V/P \qquad\qquad (4)$$

must be used, where C_V is the heat capacity of the gas (mostly air) at constant volume. Once again the ideal gas equation (2) can be used to replace the pressure P with terms involving only the volume and temperature. The result is

$$dT/T = -(R/C_V) \ dV/V$$

or, equivalently, $d(\log_e T) = (-R/C_V) \ d(\log_e V)$

Integrating,

$$\log_e(T_1/T_2) = R/C_V \ \log_e(V_2/V_1)$$

where T_1 and T_2 are the temperatures at the beginning and end of the compression, and V_1 and V_2 are the corresponding volumes.

Let's suppose our diesel engine has a compression ratio of twenty to one. For air, R/C_V is about 0.4, so $\log_e (T_1/T_2) = -0.4 \log_e(20) = -1.2$ or $T_1/T_2 = 0.3$ If the starting temperature is roughly 300 K, the final temperature is 1000 K (1340° F). This is hot enough to ignite the air-fuel mixture, which is why diesel engines do not have spark plugs. In conventional gasoline engines, the compression ratio is much smaller, typically less than ten to one, and adiabatic heating is not sufficient to ignite the air-fuel mix.

Chapter 3: Eclipses

Our family summer house in Maine is on a cliff a hundred feet above the sea. From our living room window we have a clear view of the horizon, and from a hill in back of the house, we can see for many miles in all directions. Every evening, weather permitting, we climb the hill and watch the sunset, and from our bedroom window, we can watch the moon rise. The sun sets in a wonderfully predictable pattern, a little farther to the north every day until the June solstice, and then reversing direction until it sets directly to the west at the fall equinox in late September. The height of the noonday sun varies too, and every so often I adjust the tilt angle of the solar photovoltaic collectors so we get as much electricity as we can. This is especially important when the days are short, and the lights are kept on longer.

The motion of the moon is a lot more complicated. First of all, it rises about 50 minutes later every night. Even when it is full, the place it rises varies, sometimes north ("left") of where the sun rises, sometimes south ("right"). Sometimes the full moon stays rather close to the horizon all night, and sometimes it gets quite high. The one thing that is easily predictable is that a full moon rises around the time of sunset, sets near dawn, and reaches its zenith near midnight, but the reason for this is obvious: The moon is full when the sun and the moon are on opposite sides of the earth.

Why isn't there a solar eclipse every month, at the time of the new moon, when the sun and the moon are on the same side if the earth? If the moon circled the earth in the same plane as the earth's motion around the sun, we would. On the other hand, if the moon circled the earth in the same plane as the earth's equator, we'd have two solar eclipses a year, at the equinoxes in June and September, when the sun appears to be on the equator. Wrong again! The moon circles the earth in its own plane, which is tilted about 5° from the ecliptic. So there are three planes that have to be considered, the plane of the earth's motion around the sun (the ecliptic), the tilt axis of earth's rotation with respect to this plane, which is 23.5°, and the tilt of the moon's orbital plane with respect to the other two planes. This gets complicated!

To understand all of this clearly, it is useful to view these motions from a point far outside (and north of) the solar system. If we think in this way, we can see that the earth goes around the sun in a counterclockwise direction, the earth spins once a day in a counterclockwise direction, and the moon revolves around the earth in a counterclockwise direction. All objects in the solar system turn in the same direction (except the rotation of Venus!), which perhaps offers a clue to their origin. We can think of the direction of each motion as an arrow pointing perpendicular to the plane of motion, like your right hand with fingers curled in a loose fist and thumb pointing up. These arrows all point in roughly the same direction, with the earth-spin arrow 23.5 degrees away from the earth-orbital arrow, and the moon orbital arrow 5 degrees away from the earth-orbital arrow. The earth's spin arrow, which passes through the south and north poles, points at the North Star in the little dipper.

It is a good rule in physics that "angular momentum vectors" -- that's the technical term for our arrows -- don't move much. Spinning tops (or gyroscopes) point in a direction that is fixed in space. The North Star remains the polar star. Nearly. The earth's spin axis does change a little with time, but the change, called precession (or wobble), is very slow, with a period of about 26,000 years, so for practical purposes, the North Star is the polar star during our lifetimes.

But the moon's arrow wanders. The reason is simple enough: The earth-moon gravitational interaction is not a nice neat two-body problem with a nice neat Newtonian solution. The gravitational force of the sun makes for great complexity. The forces, of course, are completely understood, and, using a computer, the motion of the moon can be predicted with perfect accuracy. But it's not the sort of calculation you can do with paper and pencil.

But it was done, eons ago, by the Chinese and the Babylonians and the Mayas and the Incas, and without the benefit of a pencil. Eclipse prediction is an ancient art. How was it done?

It isn't easy, but it isn't hopelessly difficult. What is required, mostly, is good long-term record keeping and some careful thought. The first bit of record keeping required is to time the interval between full moons. In a year, there are about twelve full moons, in ten years, about 124, in one hundred years, about 1237. The date of the full moon can be observed with an accuracy of plus or minus a day by even the most casual observer, and there are 36,525 days in one hundred years. So the period between full moons can be calculated to an accuracy of one part in 36,525, and the observed time between full moons is 29.5306 days, plus or minus 0.0008 days. This is called the Synodic month. The exceptional accuracy depends only on keeping count of the days for a hundred years or more. Discipline and political stability are what matter.

The second observation is just a little more difficult. I noted earlier that the moon sometimes rises and sets north of the sun, sometimes south; sometimes the moon is higher in the sky, sometimes lower. The maximum daily elevations of the sun and the moon are easily measured, and the maximum elevations are about the same twice a month, as the trajectories of the sun and the moon cross. The time between crossings can also be measured, and the result is 27.2122 days, plus or minus 0.0007 days. This is called the Draconic month. (It's difficult to observe the maximum elevation of the *new* moon, which occurs around noon, but it is sufficient to watch the *full* moon instead.)

For a solar eclipse to occur, the moon must be new (on the same side of the earth as the sun), and the elevation of sun and moon must be the same. This leads to a prediction: If an eclipse occurs today then another eclipse will occur in n_1 Synodic months and n_2 Draconic months, such that

$$29.5306 \, n_1 = 27.2122 \, n_2$$

All we have to do is find integers n_1 and n_2 so that the equation above is true, and this can be done by trial and error -- lots and lots of trial and error. The solution with the smallest integers is $n_1 = 47$ and $n_2 = 51$. This gives 1387.9 days (plus or minus about an hour) for the left side and 1387.8 days for the right side. This is pretty good agreement given that the discrepancy, a tenth of a day, or two and a half hours, is just about the same as the margin of error. But it isn't quite good enough. Solar eclipses are so short and so local that a two-hour error in prediction is not good enough. But the formula does work well for lunar eclipses, which last much longer and can be seen from a larger region of the earth's surface.

The next solution (aside from $n_1 = 2 \times 47$ and $n_2 = 2 \times 51$) is $n_1 = 223$ and $n_2 = 242$. This gives 6585.3 days for the left side and 6585.3 days for the right. The possible error in both values is about four hours, but the agreement is remarkable. However, there is a real problem here. What about those 0.3 days? What does it mean for an eclipse to repeat in 6585.3 days? It means the earth has turned an extra 0.3 days between eclipses. There will be an eclipse, but it will not be seen at the same place. This is a real problem: The Chinese and the Babylonians and the South Americans did not have jet planes to go chasing eclipses around the globe. A total eclipse of the sun in one place recurs as a solar eclipse thousands of miles away.

This difficulty does not occur for lunar eclipses, which cover a far wider swath of the earth's surface. So the repetition formula was not completely useless. Furthermore, the excess of 0.3 days suggests another formula, where $n_1 = 3 \times 223$ and $n_2 = 3 \times 242$. This gives 19,756 days for both sides, with a possible error of half a day. This is a very long time between eclipses -- 54 years -- but the prediction works remarkably well, better than the potential error we have estimated.

Fifty-four years is a very long time, but of course, some sort of eclipse is visible (from a fixed location) every few years. (Eclipses are visible every year from somewhere on earth, but this is irrelevant to someone who stays close to home.) Each eclipse can be counted on to repeat at an interval of fifty-four years (and thirty-four days), or, with diminished accuracy, eighteen years or 3.8 years. After one or two hundred years of observation, eclipse prediction can become very accurate, and many different eclipses can be observed and cataloged.

There is another complication that we have not mentioned. Not all solar eclipses are the same. Not only are there partial and total eclipses, but also the annular variety. The moon's orbit around the earth has an eccentricity of 5.5%, which means its distance from the earth can vary by five and a half percent plus or minus from the average. When the moon is close to the earth (perigee), we can have a true total eclipse; when the moon is far away (apogee), the maximal eclipse is annular, and only the central disk of the sun is blocked. Total eclipses are much more impressive than either the partial or annular variety.

The period between successive perigees is known as the Anomalistic month. Its length is 27.55455 days. It seems doubtful that this period was known with any accuracy by the ancient astronomers, although it is by no means impossible to measure the apparent size of the moon and chart its variation. In any case, the length of the Anomalistic month is not the same as either the Synodic or Draconic month. For an eclipse to truly repeat, the apparent size of the moon must also be the same.

A repetition period of 1387.8 days corresponds to 50.34 Anomalistic months, which is about a third of a month off true repetition. A repetition period of 6585.3 days corresponds to 238.99 Anomalistic months, an essentially perfect match for exact repetition. Likewise, a repetition period of fifty-four years also corresponds to an integral number of Anomalistic months. This is a very fortunate coincidence. In fifty-four years the sun and the moon return to the same position in the sky *and* the moon is the same distance from the earth.

Returning to our extraterrestrial position outside the solar system, we noted that the moon's orbital angular momentum vector lies close to the earth's orbital vector (the ecliptic), offset by 5°. But this doesn't specify the **direction** of the offset. The answer can be found as follows. According to the Old Farmer's Almanac, the day of the winter solstice in 1988 was December 21. On the 23rd the moon was in its "rides high" position, that is, its orbit took it the maximum angle above the earth's equator for the month, and was full. This situation is diagrammed in Figure. 3.1.

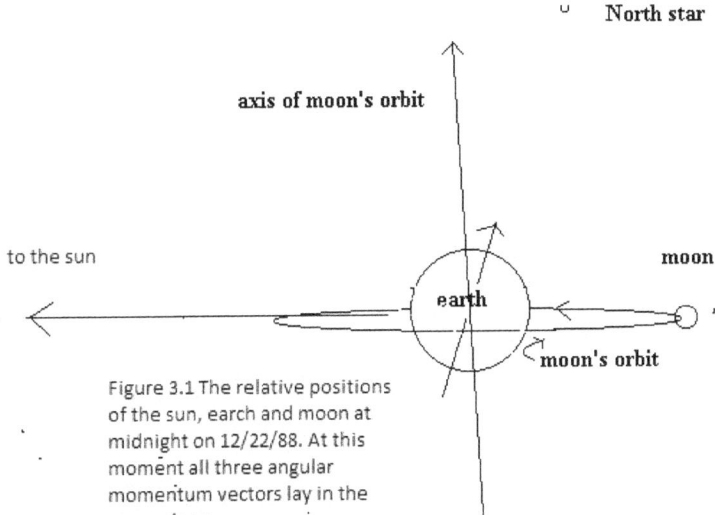

Figure 3.1 The relative positions of the sun, earch and moon at midnight on 12/22/88. At this moment all three angular momentum vectors lay in the same plane.

Ignoring the fact that the sun was just past solstice on 12/22/88, we see that the three vectors lay in the same plane, with the moon's orbital vector tilted 5° closer to the earth's orbital vector than the earth's rotation vector. In other words, the moon's orbital vector was 28.5° (23.5° +5°) away from the earth's rotation vector. At midnight in the northern hemisphere, the moon was only 90° - 28.5° = 61.5° from the North Star. The angle between the North Star and the northern horizon is equal to your latitude, so the angle between the zenith and the North Star is 90° minus

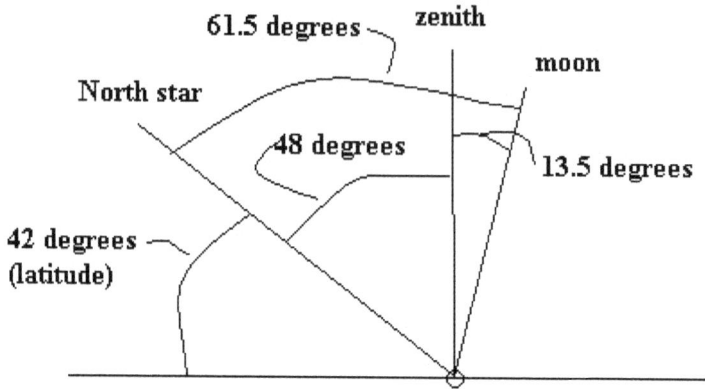

Figure 3.2 Midnight sky at Boston on 12/22/88

latitude. Thus the angle between the moon and the zenith at midnight was latitude minus 28.5°, or 13.5° here in the Boston area at 42° north latitude. Because the moon was so high in the sky, the time between moon rise and moon set was exceptionally long, nearly 16 and a half hours in Boston. This is an unusual (and simple) situation, and I chose this date because its description is especially clear. But the same sort of diagram can be constructed for other times of the year, and the reader can see that over a short time the moon's orbital vector still points in the same direction. The reader who is **very** adept at three-dimensional visualization, or reasonably adept at making models, will find that the moon's orbital vector hardly moves at all over the course of several months.

Several months, not several years! The moon's orbital vector isn't actually stationary! It moves slowly around the ecliptic (precession) with a period of 18.6 years. In three months the motion can be ignored, but in three years the change is considerable. The reader who has access to back copies of the Old Farmer's Almanac can verify this. In Table 3.1, we show how the length of the winter solstice "full-moon night" (the time between moon rise and moon set) has changed over a twenty year period in the Boston area. (At the winter solstice, the full moon is at its highest elevation of the year, just as the noon sun is at its lowest. So this is the time of year when the full moon shines for the longest time.) In 1988, this length of time was nearly 16.5 hours. Nine years later, in 1997, it was under 15 hours. As can be seen (roughly) from the table, the complete cycle takes about 18 years (18.6, to be more exact)

.The moon's position in the sky changes daily over the course of a month, from 5° above the ecliptic (the sun's position) to 5° below. Halfway in between is when solar or lunar eclipses can occur. In 1988, the moon was high and full, and low and new, in December. Ignoring the small change in precession over the course of six months, in June the moon was high and new, then low and full. No eclipses could take place during these months – the moon was 5° off the alignment of the earth and sun. Three months earlier, or nine months earlier, is when alignment occurred, and in 1988 there was a lunar eclipse on March 3, a solar eclipse on March 17, another lunar eclipse on August 27 and another solar eclipse on September 10.

Table 3.1

Length of Lunar Night (Full Moon Rise to Full Moon Set)

In the Boston Area near the Winter Solstice

Year	Time
1985	16:19
1986	16:22
1987	16:23
1988	16:27
1989	16:22
1990	16:10
1991	15:55
1992	15:37
1993	15:18
1994	15:02
1995	14:52
1996	14:50
1997	14:53
1998	15:00
1999	15:20
2000	15:34
2001	15:47
2002	16:01
2003	16:13

From the perspective of an earth-bound resident, for whom it appears that the earth is the center of the universe and the sun and the moon rise and set daily in their courses through the sky, the extraterrestrial view is hard to visualize, too hard for our ancient ancestors. Their point of view was locally based, and they didn't have accurate clocks to measure the length of time between moon rise and set. The casual modern observer has the same difficulty. The table below provides a little help in viewing the sky. The elevation of the noon sun (or the North Star) depends on latitude, and of course, early navigators used these measurements (plus a calendar) to determine their latitude on the earth's surface. The earliest maps were quite accurate in fixing the latitudes of newly discovered lands.

The sun's position in the sky can also be used to determine longitude, but only if you have an accurate clock, and clocks of the required accuracy have only been around for about two hundred years. This is why the earliest maps were so distorted. Determining latitude from the sun and the stars is easy. Until global positioning satellites were invented, longitude was hard.

The apparent sizes of the sun and the moon, viewed from the earth, are remarkably similar, almost exactly half a degree. Coupled with the precession and mild eccentricity of the moon's orbit, this means that total eclipses of the sun are rare, a once in a lifetime occurrence for most of us. In a total eclipse, the stars come out, the air temperature drops precipitously, the wind becomes still, and birds suddenly stop singing. It is a magical event that surely stimulated the imaginations and analytical skills of our ancestors, provoking a bold and ambitious attempt to understand and predict the course of nature. Science as a discipline surely had its origins in those attempts.

Problems and Exercises

Exercise #1: The full moon was only 13.5° from the zenith at midnight in Boston on 12/22/88. How far from the zenith was the full moon 9 years later?

Exercise #2: How are the Tropic of Cancer and Tropic of Capricorn defined? Over what range of latitudes is it possible for the full moon to appear at the zenith (at least occasionally)?

Exercise #3: Using a reliable data source, write down the dates of the observed **lunar** eclipses for a period of at least ten consecutive years. For each eclipse, note whether it was total, partial, or penumbral and whether it was visible from Boston. For at least the first six years of dates, add 1388 days and predict the occurrence of another eclipse. How well does this work?

Exercise #4: The diameter of the sun is 1.391×10^9 meters, while its average distance from the earth is 1.495×10^{11} meters. This means that its apparent angular size, as viewed from the earth, is 0.00930 radians, or 0.533 degrees. But the earth's orbit around the sun is eccentric ($\varepsilon = 0.0167$), so the apparent angular size of the sun varies from 0.524° to 0.542°.

The diameter of the moon is 1.391×10^6 meters, and its average distance from the earth is 3.844×10^8 meters, so its apparent size (viewed from the center of the earth!) is 0.00904 radians, or 0.518°. But the moon's orbit is much more eccentric ($\varepsilon = 0.055$), so its angular size should vary from 0.493° to 0.544°, big enough to totally block out the sun only when it is relatively close to the earth. However, there is a catch: We view the moon not from the center of the earth, but from the surface, which can be an earth's radius – 6.4 x 10^6, or 0.064×10^8 meters -- closer to the moon than the center. This increases the apparent size of the moon by 1.66%, so the angular size of the moon actually varies from 0.501° to 0.553°, although it will be smaller when it is on the horizon.

Consider a solar eclipse when the moon is at perigee, and the sun is at aphelion. How large is the region of the earth's surface that can see a total eclipse at any one moment?

Exercise #5: On October 27th, 2004, the moon was full at 11:07 pm Eastern Daylight Time. In Boston, moonrise was at 5:33 pm; it set at 7:40 am the next morning. There was also a total eclipse of the moon that night. The moon entered penumbra at 8:06 pm local time, umbra at 9:14 pm, and totality at 10:23 pm. Totality ended at 11:45 pm, umbra at 12:54 am, and penumbra at 2:03 am the next morning.

Rome (Italy) is at the same latitude as Boston (42° N) but is 83.5° to the east. What were the approximate local times for moonrise, moonset, totality, and each eclipse phase from a Roman view?

Chapter 4: Thinking About Economics

Economics is known as the dismal science, but in most respects, it isn't a science at all. Elementary texts disagree on basic principles: One text says government deficits stimulate an economy by increasing employment and enhancing demand; another says deficits reduce savings, increase interest rates and reduce investment, thereby slowing demand. Both texts cite historical examples to buttress their views, and both may well be right in certain circumstances. Economists are forever saying "on the one hand: and "on the other hand," leading to the old joke about the world needing more one-armed economists, but one thing is certain: When an economist speaks in public, his words are rarely based on scientific facts and principles. Most of what the public reads and hears about economics is political opinion dressed up as fact.

Yet another problem arises when economic statistics are reported in the news. Graphical data are often mislabeled or inadequately labeled; rates are given without a proper denominator (per month, per year); and some of the talking heads on television routinely confuse million with billion. The worst excesses and errors occur when political figures talk about economics: The mixture of bias and ignorance displayed is breathtaking and would be amusing absent the fact that political decisions are indeed economically important.

There are, of course, parts of economics that are entirely scientific, and Nobel Prizes are awarded to mathematically trained economists. Routine aspects of economics, such as mortgage interest calculations, are simply applied mathematics, and there is little disagreement about how this sort of work is done. My concern, here, is with the parts of economics where facts and politics collide. In this chapter, we will look at just one small aspect of economics, namely the deficit and debt of the United States government. A great deal of nonsense has been written about this subject, and both political parties make misleading and factually incorrect statements. I think it is instructive to look at the deficit and debt analytically, making a conscious effort to minimize political bias.

First, let us consider one way to look at federal debt by comparing it to a form of debt nearly everyone is familiar with, namely home mortgages. Roughly two-thirds of all families in the United States own their own homes, initially putting up a down payment of roughly twenty percent and borrowing the rest. For first-time buyers, the monthly mortgage payment amounts to about a third of family income, but for a fixed-rate mortgage, this fraction diminishes as incomes rise due to inflation and other factors. (First-time buyers are usually young, and their incomes also rise with increasing job experience.) Assuming a mortgage rate of 4%, this means that the size of the mortgage is about eight times income, i.e., if **I** is income, **B** the amount borrowed, and **R** the mortgage rate, then

$$I/3 = B \times R$$

Or, for **R** = 0.04,

$$B = I/(3 \times .04) = I/0.12 = 8.3 \times I$$

This ignores monthly repayment of principal, which is in any case small in the early stages of a mortgage. The suggestion has been made that government debt, like mortgage debt, should not exceed eight times government income. One fallacy in this argument is that households usually pay off at least a portion of their mortgage debt over time. More to the point, families increase and decrease in size over time; individuals grow old and die. Governments are more like corporations, and last forever, at least in theory. But eight times annual income at least provides a familiar benchmark, even if it is not quite appropriate. If perhaps you don't like a 4% rate, we can try 3%, in which case

$$B = I/(3 \times .03) = I/0.09 = 11.1 \times I$$

so that government debt might reasonably reach eleven times government income.

Nevertheless, it is of some interest to compare government and mortgage debt in another way. In 2000, the gross federal debt was 5.67 trillion dollars (5.57×10^{12}). The United States population was 285 million (285×10^6). Thus, the gross debt per capita was just under $20,000, or say $80,000 for a family of four. This is in the same ballpark as typical mortgage debt, and one can think of the federal debt as one's personal share of the mortgage the government has taken out on all the land, buildings, roads, facilities and equipment that are owned collectively by the citizenry. Similarly, when cities and towns and state governments build new schools, police stations, courthouses, libraries, roads, and other structures, these are often financed by debt (municipal bonds); interest on such debt is usually paid by local taxes and user fees. If you live in a small town, you know that new bonds are issued only rarely, when a new school has to be built due to expanding population or the retirement of an old building. Issuing debt is a way of spreading the cost of a municipal improvement over time so that no single generation of taxpayers has to pay for a facility that will be used for many years. If a town has modern facilities, it usually has some debt, and in any case interest payment on town debt is generally a small fraction of overall town

74

expenditures, which are spent mostly on the salaries and benefits of town employees: teachers, police, firefighters, and so on.

But in 2017, the gross federal debt was about 20 trillion dollars (20×10^{12}). The United States population was 325 million (325×10^6). Thus, the gross debt per capita was about \$61,500, or say \$250,000 for a family of four. So it is accurate to say that federal debt dramatically expanded in these 18 years.

Federal debt is different from personal and municipal debt in a variety of ways. First, federal debt is not tied to specific buildings and structures but can be issued for any reason. Second, the federal government can, to a certain extent, manipulate the value of the currency through the actions of the central bank. Third, the federal government can alter the tax code, changing the amount of taxes collected. (States and municipalities can also change their tax rates, but the effects are dispersed over many thousands of separate governments.)

In many ways, federal debt is similar to corporate debt. Corporations, like governments, have an unlimited lifetime; their budgets grow (or more rarely, shrink) over time, and both issue bonds to finance routine activities. One crucial difference is that corporations raise funds in ways quite different from governments, through sales of goods and services, borrowing (debt), or the issuance of equity (stocks). Governments generally do not charge for the goods and services they provide except as user fees, but instead, collect taxes. As for equity, the government is owned equally by all citizens, and in any case, does not account for its holdings of goods and property. So the analogy between corporations and the federal government is a little weak, although lenders to both institutions take the analogy seriously because a lender with cash on hand can choose where to lend money. A company that has solid profits and little debt (compared to equity) is an excellent credit risk and hence will pay a low rate of interest. A company with a higher debt-to-equity ratio will pay a higher rate of interest since the risk of default is presumably greater. Loans to the federal government always have the lowest interest rate since the risk of default is essentially zero, or so investors believe.

With this background in mind, we can now take a look at the federal budget. In 2000, government income was 2.03 trillion dollars. Expenditures were somewhat less, so there was a surplus of 0.167 trillion (i.e., 167 billion), and the "debt held by the public" decreased from 3.64 trillion to 3.41 trillion (with a small error due to other factors). However, there is another way of accounting for the debt, the "gross federal debt," which eliminates internal government borrowing, largely borrowing from the social security trust fund to finance other government activities. The gross debt in 2000 was 5.68 trillion, having increased from 5.66 trillion the year before. By this measure, the federal government ran a deficit of 0.02 trillion (i.e., 20 billion) for the year.

Compare that with 2015. Government income was 3.25 trillion, but expenditures had grown to 3.69 trillion, leaving a deficit of 0.439 trillion (439 billion), and the debt held by the public increased from 12.8 trillion to 13.1. Likewise, the gross debt increased from 17.79 trillion to 18.12 trillion. The federal budget was now far from surplus.

The question we are trying to wrestle with is whether these numbers are in some sense good, bad, or indifferent. One way of answering this question is to look at the size of government interest payments. In 2000, federal interest payments "to the public," that is not including payments to the social security administration, were 0.223 trillion dollars (223 billion), or 11% of tax receipts. This was the lowest percentage in two decades. Furthermore, in the years between 2000 and 2017, the percentage remained close to 8%. Since 1950 the fraction has varied from just under 7% to just over 18%. (See Figure 4.1) Interest payments were a relatively small fraction in 2000 because (a) inflation and interest rates were lower than in previous years and (b) the federal budget was running a surplus (or at worst a relatively small deficit) while tax receipts were increasing, thus reducing federal debt as a percentage of receipts. Is 8% in some sense "better" than 18%? It is hard to say, although it is clear that 18% is in some sense dangerously high. At 100% the government would be bankrupt, with all tax receipts used to pay interest, and even at lower numbers an increasing percentage hints at poor fiscal management. (Nevertheless, 18% is well under the 33% that a young family might pay in mortgage interest on a new home.) However, interest payments as a fraction of federal receipts are a relatively volatile measure, and it is worth looking for measures that are more stable, less subject to sharp swings in interest rates.

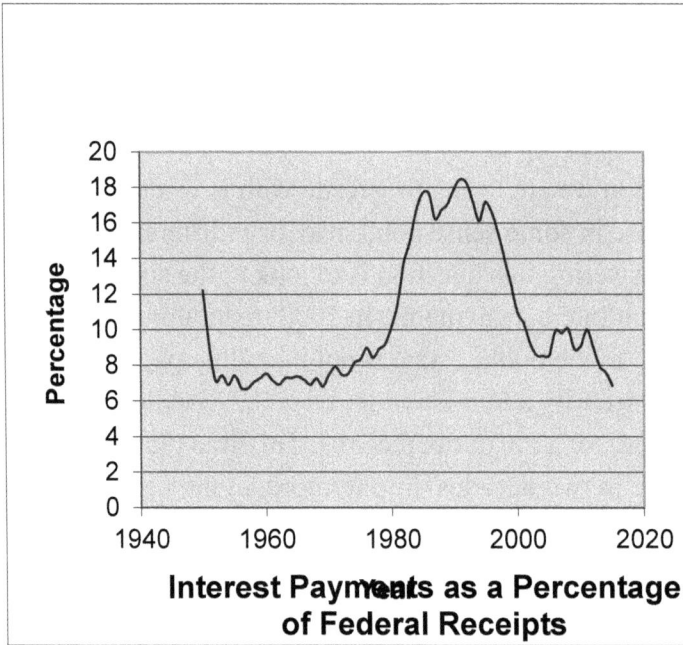

Interest Payments as a Percentage of Federal Receipts

Figure 4.1

(See also Exercise #7)

One such measure is the ratio of government debt (or expenditures) to the gross domestic product (GDP), the total output of goods and services of the United States economy. This would be an odd measure for a corporation – debt as a percentage of sales – but for the federal government, GDP is probably the best measure of the revenue base. Figure 4.2 shows both sets of numbers starting in 1950.

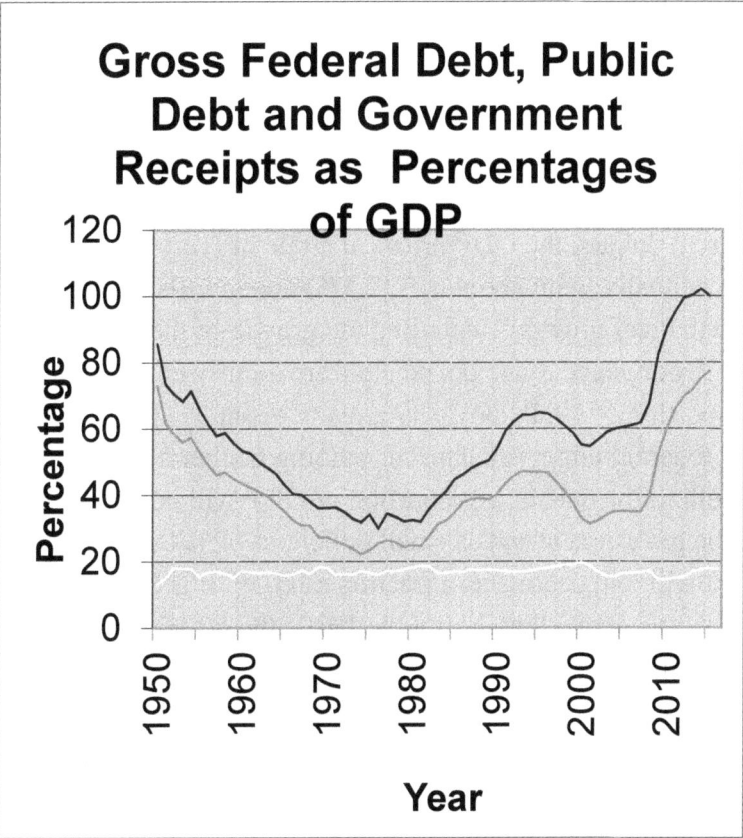

Gross Federal Debt, Public Debt and Government Receipts as Percentages of GDP

Figure 4.2

The figure shows that tax receipts have remained fairly constant over the past sixty-five years, ranging from 13.1% of GDP (1950) to 19.8% (2000), with the trend recently slightly upward. Debt, on the other hand, has varied enormously, changing by over a factor of two. The variation is not as great as interest payments as a percentage of government receipts, but it is still a large variation.

It turns out that the government does not need to run a surplus in order to reduce the effective size of its debt. In current dollars, the GDP grows at a rate of (very roughly) 5% annually, with about half (2.5%) due to inflation and half due to "real growth," including the growth of the population (~1% per year). Since tax receipts are a fairly constant proportion of GDP, they also grow at about 5% per year. If the federal budget is balanced, running neither a surplus nor a deficit, the debt as a proportion of GDP will decline. Even if the government runs a small deficit, say 2% of GDP, the debt will still decline as a fraction of GDP. It is clear from the figure above that during the 1980s the government ran unusually high deficits, which is puzzling since the nation was not engaged in any major war nor was the economy in particularly grave difficulty. The same is not true for the decade starting roughly in 2008 when there was a major financial crisis.

We have not yet dealt with the question as to **which** debt is of primary concern, the gross debt or the public debt. In 2000, the social security system collected 653 billion dollars in taxes on employers and employees. Expenditures, however, were considerably less, 409 billion. The surplus was lent to the rest of the federal government, which (for once!) did not spend it all, but instead used most of it to reduce the debt to the public. However, at the end of the year, the federal government owed the social security system 2.27 trillion dollars, the difference between "the gross federal debt" and "the debt held by the public." This seems huge until one realizes that, for actuarial reasons (decreasing birth and death rates; aging of the population) the social security administration has had to run a surplus. The amount of that surplus was determined largely by actuarial accounting, not by politics. What was to be done with the extra money? The safest place to lend it, surely, was the federal government.

It should be noted, parenthetically, that some politicians, mostly Republicans, have suggested that a portion of social security payments should be invested in equities, i.e., stocks. However, social security is not a complete retirement system but provides only about 40% of the average worker's pre-retirement salary. Retirees need to supplement social security with other income, such as a corporate pension, a 401(k) plan, personal savings, or cash from the sale of a house.

Following the principle that retirement funds should be diversified, it seems reasonable that the social security portion should be the safe, guaranteed segment. Workers themselves need to save or invest in equities in addition to social security in order to assure a comfortable retirement. Nevertheless, this does raise the question as to whether the social security administration should lend its entire surplus to the federal government. A more prudent plan might be to mix in some corporate lending and thus achieve a higher rate of return. However, I am not aware of any politician having ever made such a suggestion. The federal government itself often makes loans to individuals, corporations, and other governments, assuming by itself the risk of default and bankruptcy; private banks do the same. The social security administration could also do so with minimal risk.

As some point in the near future, around 2020, the social security system will no longer run a surplus and will begin receiving repayment from the federal government. What then? Some sort of tax increase seems inevitable. How this will be done is impossible to predict, but almost surely it **will** be done, because the alternative is to cheat the people whose money has been used to fund the social security system via a flat tax with no deductions, a tax that would be severely regressive absent the assurance that payments are really a form of enforced savings. My own guess is that some form of value added tax (i.e., a sales or consumption tax) will be used to raise the additional funds, as is done in the European community and in most other wealthy nations of the world.

The alternative, the Japanese model, is to try to drive down interest rates so low that interest payments on the debt are small. This is, in a way, bankruptcy: There is very little difference between paying historically minimal interest and paying no interest at all.

In the interim, it would seem prudent to keep the "debt held by the public" to a constant or declining percentage of the GDP. What is to be avoided, above all, is a sudden shock to the U.S. economic system caused by a sudden large increase in taxes or debt, or a sudden major change in the way taxes are collected. Economic systems generally respond well to gradual change, but far less well to changes that are abrupt or unforeseen.

In making this analysis, I have attempted to steer well away from political considerations and to examine the debt/deficit questions in an analytical way. Scientific objectivity in political matters is clearly impossible, but an honest search for objectivity is surely worthwhile. This is how a scientist is supposed to think, and how I think readers can inform themselves on economic matters. The data are readily available on government and private websites. And if it is politics that are of greater interest, one can always correlate the swings in the economic indicators with the political party in power; the results surely have something to say about fiscal "conservatives" and "liberals."

Problems and Exercises

Exercise #1. For each of the other six G7 nations, obtain the following current data: GDP, population, government receipts as a fraction of GDP, government debt, and interest paid on such debt. Compare appropriate ratios with those of the United States.

Exercise #2. Select a single G7 nation and examine how ratios similar to those shown in Figure 4.1 and Figure 4.2 change over time.

Exercise #3. If the federal government ran a balanced budget every year, how long would it take for the gross federal debt to drop from 100% of GDP to one-third, assuming an annual GDP growth rate of 5%?

Exercise #4. Prepare a graph showing federal expenditures as a percentage of GDP over time. On the same graph, show federal receipts.

Exercise #5. We have omitted state and municipal debt and interest in putting together the data for this chapter. What percentage of GDP is total governmental debt in the United States? What percentage of GDP is total governmental revenue?

Exercise #6 Another important economic concern is the trade deficit, or, more properly, the accumulated trade imbalance. What percentage of GDP is the current trade deficit? What percentage of GDP is the accumulated trade imbalance? What happens to the dollars that flow to other countries as a result of trade imbalance? Is GDP a suitable denominator when examining trade issues?

Exercise #7. Bring Figure 4.1 up to date. What is happening? What changed since this text was written?

Part II: Health and Risk

Chapter 5: Photons, Cell Phones, Sunlight, and X-Rays

Electric power is, without question, one of the greatest inventions of the human species. Just think: You plug your gizmo into the wall socket, and outcomes all the power you need, for as long as you need it. It's cheap, too: An hour's use of the electric company's horse[8] costs less than a dime, and you don't have to feed it or clean up afterward. Electric power is one of the miracles we take for granted, and we notice it only when it is taken away, during those rare days of blackouts and violent storms. Electric power cools and heats your home, washes and dries your clothes and dishes, fires up your computer and television, and lights up your nights. It is, as a spokesman for the industry once said, women's liberation, although he should have mentioned the birth control pill in the same breath.

Electric power isn't entirely safe, of course. In the United States, about 1000 people die every year from accidental electrocution. That's two or three in a million, roughly speaking, a small enough number that hardly anyone notices, except for the friends and kin of the deceased.

---

[8] A horsepower is 745 watts, so a horsepower hour is 0.745 kilowatt hours.

There are other hazards of living with electric power. Coal-fired power plants emit acids and particulates (dust), nuclear plants have radiation dangers, and hydroelectric plants despoil the landscape. These hazards are harder to quantify, although they are all surely real, if in some cases perhaps small. We are all guinea pigs in the progress of technology, although the good news is that, on balance, our life spans keep increasing, year after year.

One danger of electric power is almost surely not real – the presumed hazard of living under high voltage power lines, or, for that matter, sleeping under electric blankets. For a time it was widely believed that close proximity to electric power lines could cause a variety of health problems, including childhood leukemia and adult cancer. The belief originated in a statistical association. After much study, the association appeared to weaken and has since been attributed to other causes. Wealthy neighborhoods can usually fight off having power lines placed locally; poor neighborhoods are not so successful. Poor people have more health problems.

More recently, an alarm was sounded over the use of cell phones. Maybe cell phones cause brain cancer. The alarm didn't last very long. I think people found cell phones so convenient they were willing to ignore any unproven risk. The perceived risk in high-voltage power lines was economic: Unsightly power lines reduce property values. The perceived risk of cell phones seems limited to the annoyance of other people's use, and the hazards of driving while dialing or talking or texting.

To understand why most scientists think power lines and cell phones do not pose direct risks of health problems requires a little background in electromagnetic radiation and quantum mechanics. All radiation, from radio and television to visible light and x-rays, consists of particles of energy called photons. All photons move at the speed of light but differ in the amount of energy each photon carries. Radio waves carry the least energy per photon, gamma rays, and x-rays carry the most, and visible light is in between, blue light being more energetic, red light less so. Photons are absorbed by chemical matter in different ways depending on their energy: x-rays smash into matter and break chemical bonds, while radio waves have very mild effects that involve heating up whole molecules and making them move a little faster. An intense beam of radio waves cannot break chemical bonds, while even a weak beam of x-rays can. (So can the ultra-violet portion of sunlight.) This is why nuclear radiation, which consists of photons (and other particles) even more energetic than x-rays, is dangerous.

The "flux" of energy radiation can be stated in a couple of different ways. The common measure is watts per square meter of surface area (or watts per square centimeter), and a good standard is the sun's radiation on the earth, about 1000 watts per square meter, or a tenth of a watt per square centimeter. This is about the same energy flux your skin would feel if you held your hand nine centimeters from a 100-watt incandescent light bulb. It's warm but safe. The other measure of flux is the number of photons per square meter (or square centimeter) per second. For a standard of 0.1 watts per square centimeter, this varies enormously depending on whether each photon carries a lot of energy or a little. For radio and TV and cell phones, there are a lot more photons in 0.1 watts per square centimeter than in an equal beam of x-rays, but the x-rays would do serious damage to your hand, while a radio wave would probably do little more than heat your skin. It's not the number of photons that matters most, but the energy per photon.

Living cells and organisms have repair mechanisms, and it seems likely that the human body can cope with low levels of nuclear radiation, just as it can cope with a mild sunburn. For example, people who live at high altitudes, such as Colorado, are exposed to far higher levels of cosmic rays (a kind of super-energetic x-ray) than people who live at sea level, yet do not have higher cancer rates. It seems that there is a safe level of exposure to cosmic rays, below which no harm can be seen. On the other hand, people who spend a lot of time outdoors, exposed to the sun, have far higher rates of skin cancer. Photons of ultra-violet radiation from the sun have enough energy to break chemical bonds, cause real damage, and overwhelm the biological repair systems. When it comes to cell phones and electromagnetic radiation from power transmission lines and electric blankets, most scientists assume that it is extremely unlikely that the low energy photons involved can have any effect at all as long as the overall intensity of the radiation is below the level where strong heating can occur (as in a microwave oven). As long as the overall energy level is below that of sunlight, it is difficult to see how such radiation can cause any harm. The photons that originate in a source like a cell phone are far too weak to break chemical bonds.

We can imagine some mechanisms by which damage could be done. After all, a cell phone emits nearly a watt, so parts of your skin will be exposed to energy levels around 0.1 watts per square centimeter. And this level is not always safe: Looking at the sun for more than a few seconds can do permanent damage to your retina. Perhaps cell phone radiation passes through your skull and is preferentially absorbed by a small section of your brain, which is then heated and cooked. This is not at first glance an absurdly silly hypothesis, but there isn't enough energy emitted from a cell phone to do much cooking. One can try to invent other mechanisms, but it is hard to overcome the basic facts that the overall energy level is small and the photon energies too small to break bonds.

Could scientists be wrong about this? Yes. But I bet my life they aren't. So have tens of millions of fellow guinea pigs.

Here's a much more detailed and technical way of talking about photons. The energy of a single photon is given by the formula $E = h\nu$, where $\nu = c/\lambda$. (Energy equals Planck's constant times frequency; frequency equals the speed of light divided by wavelength). For example, from the second formula, we can calculate that the wavelength of a cell phone operating on the 850-megahertz band ($\nu = 850,000,000$ hertz) is about a foot (35 centimeters, to be more exact) since the speed of light is 30,000,000,000 centimeters per second. From the first formula, the energy of a single photon from a cell phone antenna is 5.6×10^{-25} joules. A watt is a joule per second, so a one-watt phone emits 1.8×10^{24} photons per second. That's a lot of photons!

For comparison, visible light has wavelengths around 5×10^{-5} centimeters, so the energy of a single photon is around 4×10^{-19} joules, far more than a photon from a cell phone. Correspondingly, the number of photons is much smaller, and our standard solar flux of 0.1 watts per centimeter squared is equivalent to 2.5×10^{17} photons per square centimeter per second. It is not surprising that looking directly at the sun can damage your retina given that the eye is able to see light as dim as a few dozen photons in a short flash. The central part of the human retina is only 0.01 square centimeters in size (a square millimeter) and contains about 150,000 cells – about a third as many pixels as your (much larger!) computer screen. So looking at the sun means that each cell is receiving billions of photons per second.

Consider, by contrast, the light from a candle, which emits about 0.0015 watts in the visible part of the spectrum. At a distance of a meter from the candle, the light intensity is about 1.2×10^{-8} watts per square centimeter. At ten meters, 1.2×10^{-10}. At one hundred meters, 1.2×10^{-12} watts per square centimeter. At this distance, you can still see the candle, but your eye is receiving only three million photons per square centimeter per second. Since the diameter of your pupil is only about 6 millimeters (dark adapted), only about a million photons per second enter your eye, which has a focal length of about 1.7 centimeters, roughly f/3 in camera terms. The image of a candle one centimeter in diameter at a distance of 100 meters (10^4 centimeters) has a diameter of only 1.7×10^{-4} centimeters on your retina, or an area of 2.3×10^{-8} square centimeters. This is less than the size of a single receptor cell! But your eye, by doing some clever movements, and some equally clever computing, can still see the candle, which is, in effect, a point source, like a star in the sky. The eye can see stars with a magnitude of 6 or less, corresponding to an intensity of about 10^{-14} watts per square centimeter, or about 10,000 photons per second on your retina. And you could do even better with a flickering light source. As the figures below indicate, the eye can deal with an extraordinary range of intensities although, outside the range 400 to 700 nanometers, it has no sensitivity at all.

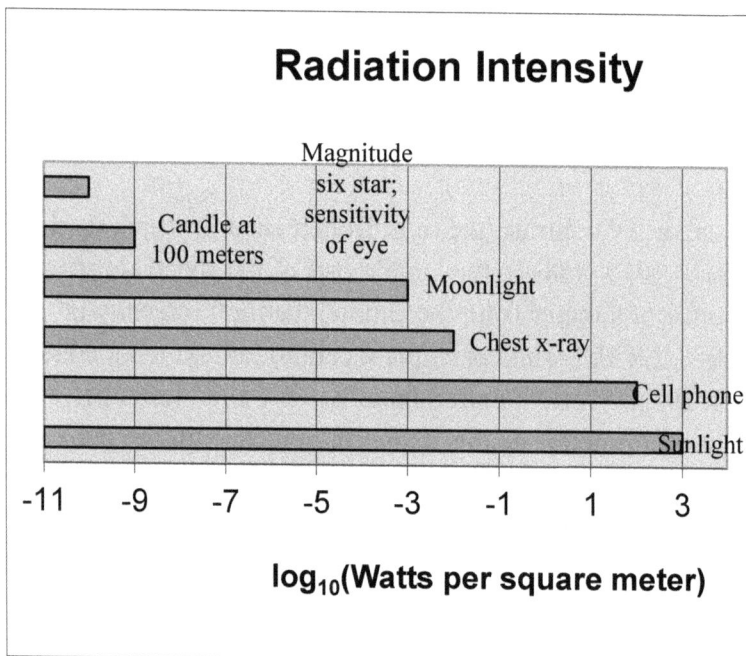

Figure 5.1 Intensity of Various forms of Radiation, log(10) Scale

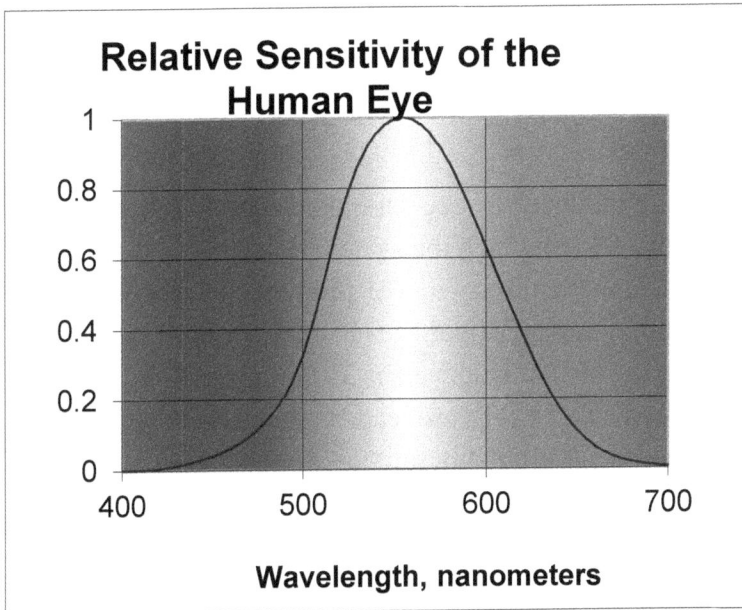

Figure 5.2 Relative Light Sensitivities of the Human Eye

Visual acuity is another matter. The eye can only resolve two points that are farther apart than about a tenth of the diameter of the moon. This is equivalent to two objects a centimeter apart a thousand centimeters away. On your retina, these two points would be 1.7×10^{-3} centimeters apart. A circular spot of this diameter would include about 30 receptor cells. This is why we cannot see the mountains of the moon without a telescope.

We have seen that the eye can deal with an extraordinary range of visible electromagnetic radiation, from 10^{-14} to (perhaps) 10^{-2} watts per square centimeter. What about x-rays?

X-rays are produced by accelerating electrons in a cathode ray tube (CRT), in much the same fashion as electrons are accelerated in the tube of an old-fashioned television set or computer monitor. The voltages in an x-ray machine are a little higher, 60,000 volts versus 35,000 in a TV, but of course, consumer CRTs use lead crystal glass to reduce x-ray emissions to a presumably safe level, while x-ray machines are designed to convert the energy of the electrons to x-ray photons. Due to losses in the conversion process, the average energy of an x-ray photon is about 30,000 "electron volts," which corresponds to a wavelength of about 4×10^{-9} centimeters. X-ray wavelengths are extremely short and are usually measured in Angstroms or nanometers or picometers. (One Angstrom is a tenth of a nanometer or 10^{-8} centimeters. A picometer is 10^{-12} meters or 10^{-10} centimeters.) An x-ray photon with a wavelength of 40 picometers has an energy of 5×10^{-15} joules. A flux of 0.1 watts per centimeter squared is equivalent to about 2×10^{13} photons per square centimeter per second. Actual x-ray exposures are far less intense, and the exposure lasts only a couple of seconds. What matters here is not the flux, but the total dose. A typical chest x-ray delivers a dose of 0.1 mSv (milliSieverts), where a Sievert[9] is defined as a joule of photon energy absorbed per kilogram of body mass. Assuming that your chest region weighs about ten kilograms, it absorbs about 0.001 joules of radiation during an x-ray. By way of comparison, a mere one-second exposure of your chest area (about 1000 square centimeters) to sunlight would deliver 100 joules. So a chest x-ray doesn't sound like much radiation until you recall the difference in energy per photon. A dose of 10 Sieverts can kill you.

Photons per Joule

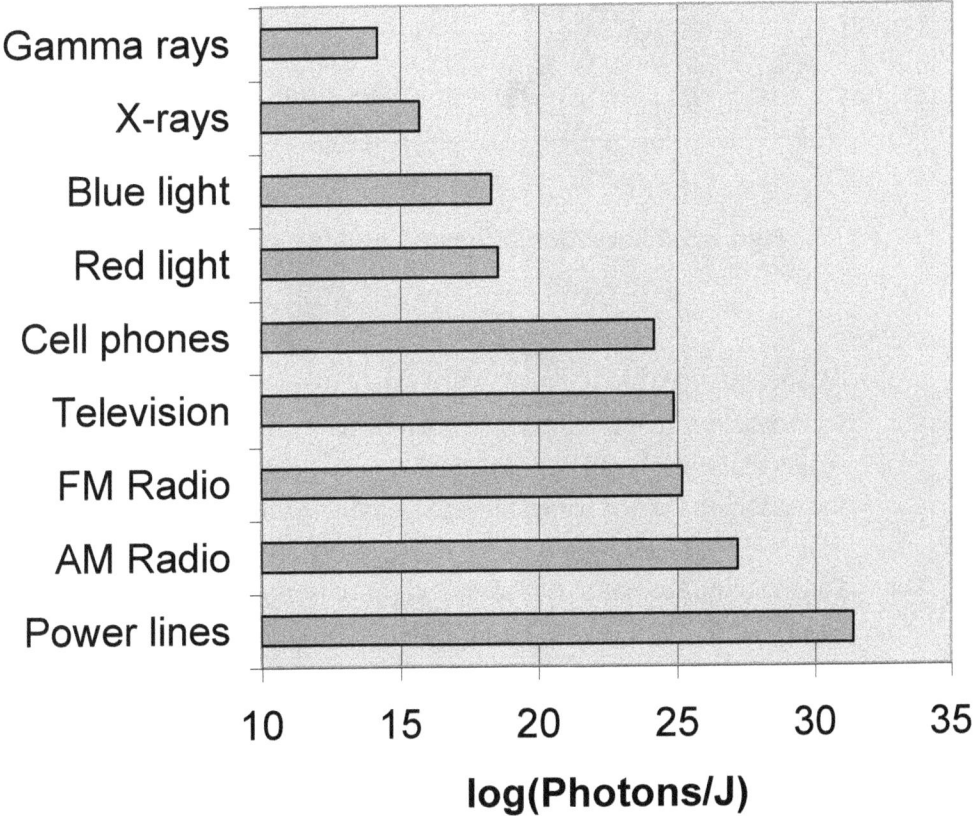

Category	log(Photons/J)
Gamma rays	~14
X-rays	~15.5
Blue light	~18
Red light	~18.5
Cell phones	~24
Television	~25
FM Radio	~25
AM Radio	~27
Power lines	~31

log(Photons/J)

Joules per Photon

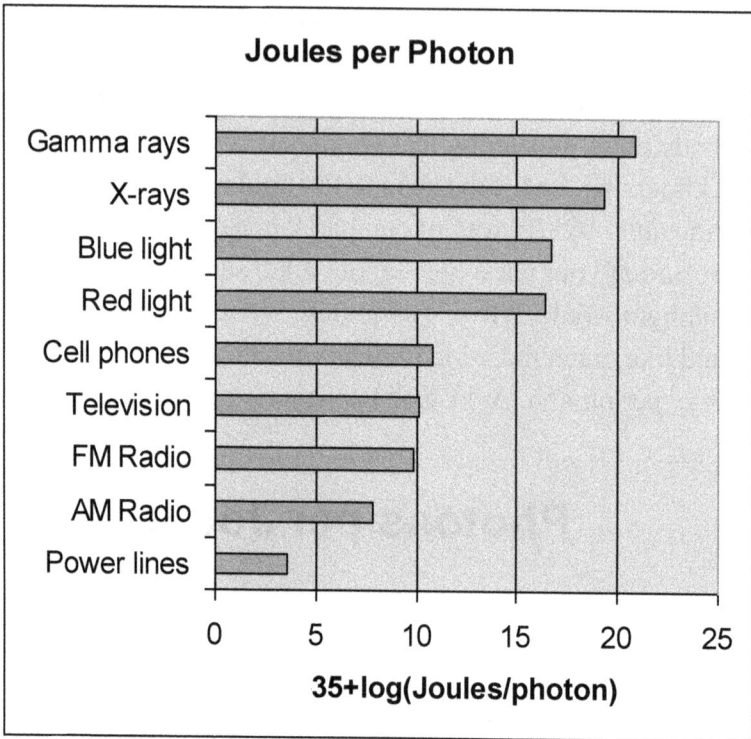

Figure 5.4 Radiation Intensity, Joules per Photon

To understand what happens when the body is exposed to x-rays, consider that a ten Sievert dose to your chest area means that your body has absorbed 2×10^{16} photons, each one of which causes some damage to a chemical bond in your body. Ten kilograms of your body (which is mostly water) contains about 10^{27} atoms. So only five atoms in 10^{10} are affected – five atoms in ten billion. Not many, surely, but enough to kill. Yet a dose one hundred thousand times smaller is presumably[10] safe. Just why the disruption of only five atoms in ten billion can kill is not well understood, nor is it clear why smaller doses are safe. Biological systems have mechanisms for repairing cell damage or destroying damaged cells, and, presumably, as long as these mechanisms can keep up with the rate of damage, full and rapid restoration is likely. Once these mechanisms are overwhelmed, repair is no longer possible. A cell[11] contains (very roughly) 1.5×10^{14} atoms, so a ten Sievert chest dose affects three cells in ten thousand. A dose of 0.1 mSv affects only three cells in a billion. That seems to be a level of damage your body can cope with.

What we have shown here is that the risk from radiation has little to do with overall intensity (as long as the intensity is insufficient to cook you). What matters is the way radiation and atoms interact. Weak photons from cell phones are probably harmless; stronger photons from visible light also cause little harm although the eye can detect amazingly low levels of energy. But when it comes to the highly energetic photons in x-rays and cosmic rays, very low doses of energy can destroy cells and kill, although it is likely that as long as only a small fraction of cells are killed, the body can cope.

---

[10] We will examine this presumption more carefully in the next chapter.
[11] There are about 675 million cells in a cubic centimeter.

Problems and Exercises

Exercise #1. If cells were uniform spheres, packed together like oranges in a supermarket, they would occupy 74% of a given volume, with the rest taken up by the intercellular fluid. Since there are approximately 675 million cells per cubic centimeter, such cells would have a diameter of 1.28×10^{-3} centimeters. **A:** Verify this calculation. **B:** If cells were uniform cubes, with no intercellular fluid, show that the length of a side would be 1.14×10^{-3} cm. **C:** If a uniform layer of cells had a thickness of just one cell, how many cells would there be in an area of 0.01 square centimeters? How does this compare with the density of cells in the central part of the retina? Why are these cells called rods?

Exercise #2. The average energy of an x-ray photon is about 30,000 electron-volts (eV), while the energy required to break a typical chemical bond is only about 3 eV. In principle, there is enough energy in an x-ray photon to break 10,000 bonds. It is generally believed that the damage caused when an x-ray photon interacts with a cell is extremely localized, confined to the immediate vicinity of the collision site so that the interaction affects only one (or at most two) cells. This means that a 10 Sievert chest dose affects far more than five atoms in ten billion while still disrupting three cells in ten thousand. Compare the expected effects of a 10 Sievert dose of x-rays with the same dose of gamma radiation. Which dose damages more cells?

Exercise #3. In a photoelectric cell, a voltage and current are produced when the energy per photon of light on the cell is greater than a critical value. Weaker photons produce no effect, while more energetic photons produce one electron per photon at the critical voltage. (This is called the photoelectric effect; it was first explained by Einstein in 1905.) A modern solar panel consists of 36 silicon photoelectric cells wired in series, with a total open-circuit voltage of 21 volts, i.e., 0.58 volts per cell. What is the minimum energy per photon required to produce this voltage? What is the maximum effective wavelength? How goes this compare to the wavelength of yellow light, 5×10^{-5} cm? How does this affect the efficiency of a silicon solar cell?

Exercise #4. The pixels in a digital camera are basically tiny silicon photoelectric cells. Each absorbed photon produces an electron, and the electrons are stored and later counted to produce an image. Since each absorbed photon produces only one electron, how is it possible to produce a colored image?

Exercise #5. The cells in your retina act in a way that is very similar to the pixels in a digital camera, except that photon absorption produces a reversible chemical change rather than an emitted electron. How is it possible for us to see colors?

103

Chapter 6: The Linear Dose No Threshold (LNT) Hypothesis

In the previous chapter, we noted at a ten Sievert dose of x-rays is likely to kill you (within days or weeks), while a dose of 0.1 mSv (a typical chest x-ray) is regarded as safe. Suppose, however, that a 0.1-mSv dose isn't safe, but kills one in every one hundred thousand people, the ratio of the two doses. Or, alternatively, suppose that since a lethal dose takes (on average!) forty years off the lifespan of the recipient, assume that a 0.1-mSv dose takes 40/100,000 years (a few hours) off the life of each person who receives an x-ray. How can we tell which hypothesis is true?

We can't – not with any confidence. There is no obvious way to distinguish among these hypotheses. A death rate of one in one hundred thousand sounds like a lot, but it isn't. The death rate among girls age 5 to 14 – the lowest rate among any group of Americans – is 15 per hundred thousand per year, and any child of that age who gets a chest x-ray presumably has other health problems that would confound any analysis. The only tests that could reliably distinguish among hypotheses are experiments only Nazis would think of performing.

There's a fourth hypothesis worth considering: A chest x-ray might increase your lifespan, and not just in the obvious way – doctors order x-rays to diagnose and treat what ails you. Some scientists believe that an x-ray might stimulate your cell repair mechanisms, much as a vaccination stimulates your immune system. Low x-ray exposures might actually be beneficial. People who live at high elevations, where they receive a bigger dose of gamma rays than those who live at sea level, actually live longer! So this fourth hypothesis is not without some evidence, and cannot be dismissed out of hand.

The linear dose hypothesis[12] is an assumption widely used by regulatory groups and agencies. The assumption basically states that there is no such thing as a safe dose, that the hazard of a dose is proportional to the size of the dose. Chest x-rays, according to this hypothesis, are not safe, but the benefits are presumed to be more than sufficient to warrant their use. The problem, of course, it that the hypothesis is nearly impossible to prove or disprove, and may well be too conservative. Or too risky.

Here are some hypothetical examples of risk-dose relationships. Figure 6.1 shows the LNT equation, in which risk is proportional to dose, and no safe dose exists. Figure

---

[12] Also known as the "linear dose no threshold" assumption, or LNT hypothesis. "No threshold" means that the risk-dose curve is not only linear, but goes through the zero-dose, zero risk point on the graph.

6.2 shows a linear dose relationship in which there is a safe cutoff dose below which the risk is zero. Figure 6.3 shows a supra-linear relationship, in which low doses are disproportionately riskier than large doses.

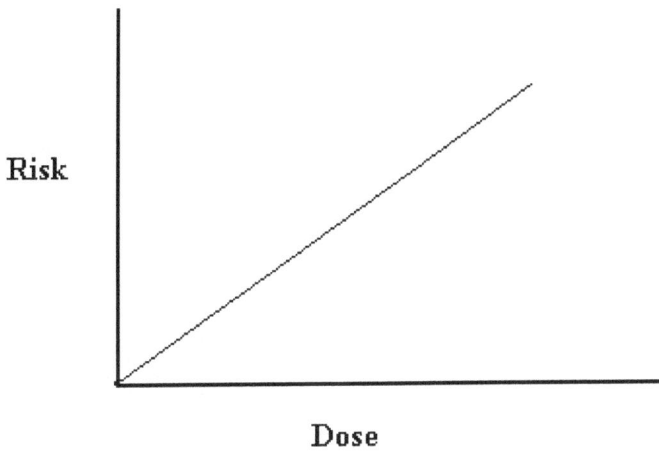

Dose

Figure 6.1 The Classical Linear Dose No Threshold Plot (LNT)

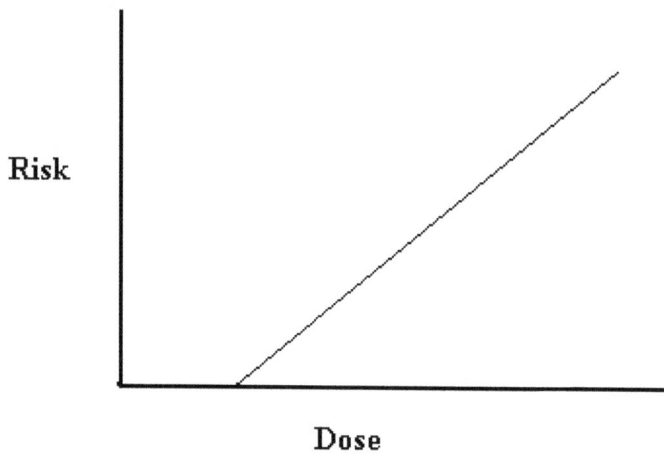

Figure 6.2 Linear Dose with a Safe Threshold

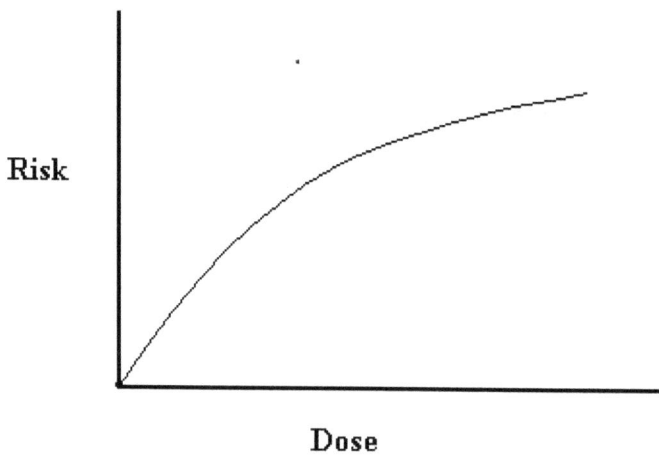

Figure 6.3 Supra-Linear Relationship

Is the linear dose hypothesis too risky when it comes to radiation and x-rays? That is, is the risk of large (but sub-lethal) doses greater than the prediction? From time to time, people are accidentally exposed to large amounts of radiation, and it does seem that the dose-risk equation is at least roughly linear when the doses are close to the lethal level. For obvious reasons, however, the data are scanty and hard to quantify, and the best that can be said is that the linear dose hypothesis does not seem to underestimate risk. But at very low doses it is hard to distinguish the linear curve from the J-curve, in which very low doses actually produce a benefit.

There is one fairly well-understood case of a J-curve risk-dose relation: Alcohol consumption. Alcohol, or more technically, ethanol or ethyl alcohol, is an unquestioned poison. Consumption of half a liter of alcohol (roughly 15 ounces, or 30 "drinks") over a short period of time (hours) will kill you. Period. Yet there is overwhelming evidence that regular consumption of one ounce a day will prolong your life by several years. Figure 6.4 shows some of the best available data[13]. The "risk" side of the chart is expressed as "death rate," not years of life expectancy, so the numbers may be hard to grasp. For comparison, the overall annual death rate of men age 55-64 is 1,254 per hundred thousand, while the rate for women in the same age range is 1,005 (data for the year 2000). So a decrease in rate by 250 in

---

[13] Data from New England Journal of Medicine, Volume 337, p.1705 (1997).

100,000 is roughly equivalent to the life expectancy difference between men and women, about 6 years. Moderate alcohol consumption is, without question, good for you.

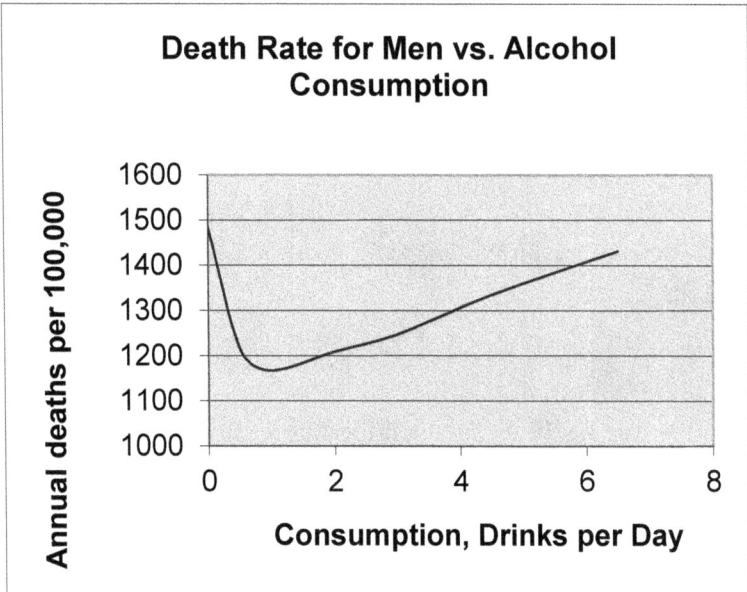

Death Rate for Men vs. Alcohol Consumption

Figure 6.4 The J-Curve for Alcohol Consumption

Data like those presented in Figure 6.4 are not easy to collect. It is hard to interview dead people about their drinking habits. To collect data of this type, tens or hundreds of thousands of people must be studied, and then followed for years until they die. Such "longitudinal" studies are expensive, and require consistency in funding. A short gap in data collection or patient tracking can ruin a study.

The deleterious effects of alcohol are well understood: Drinking reduces inhibitions, decreases mental performance, and harms the liver. In the United States, about 40,000 people die each year in motor vehicle accidents, which is equivalent to a death rate of about 14 per 100,000. Many of these deaths are alcohol-related. But the risk is dwarfed by the benefits.

Why is alcohol beneficial? Much of the benefit is clearly attributable to effects on the heart and cardiovascular system. Alcohol appears to raise the level of HDL ("good") cholesterol in the bloodstream, reducing the chances of a heart attack. Since heart disease is the number one cause of death, at least part of the linkage is clear. However, this is not the whole story. The rate of cardiovascular death as a function of alcohol consumption has been studied, and while alcohol is clearly beneficial, the benefits to the heart alone are inadequate to explain the overall drop in death rate. We really don't know in full detail why alcohol is good for us. Maybe it's because of the antioxidants in wine (and grape juice!), maybe wine is better than scotch. But we don't know. The relation between diet and life expectancy is a subject of great interest and even greater uncertainty, and only very long-term studies involving hundreds of thousands of people will provide answers.

There is a temptation to shortcut such studies by using "proxies." For example, higher HDL levels can be a substitute or proxy for a reduction in overall death rate. It is obviously much easier, faster and cheaper to measure HDL levels than to track people over the courses of their lives. In the case of alcohol consumption, HDL levels are a seductively bad proxy, as is the death rate due to cardiovascular disease. The more you drink, the less likely you are to die of cardiovascular disease, even at the highest levels of measured consumption, six or more drinks a day.

In the case of alcohol consumption, there are hundreds of millions of people who are potential participants in any study of risk. For most other hazards, such as radiation, the risk pool is much smaller, and the studies are correspondingly more difficult and less accurate. The smaller the sample size, the greater the chance that the results are mostly statistical noise. Most data come from accidents, especially industrial accidents and exposures to unusual chemicals and circumstances. For example, most sailors are familiar with the 50-50-50 rule, which states that a person in the water for fifty minutes at a temperature of 50°F has a 50% chance of survival. But the most basic ethical principles forbid us from doing experiments that would deliberately expose human beings to hazards. There are, of course, animal studies, but extrapolation to humans has its own set of uncertainties.

However, in wartime, ordinary ethics are often cast aside, and sometimes horrendous experiments are done, accidentally or on purpose. I am not thinking just of the experiments done by Nazi doctors during World War II. The bombings of Hiroshima and Nagasaki exposed over a hundred thousand people to radiation. Many died within days or weeks; people farther from the blast epicenter often died within years. But for people farther away, there is good evidence for a J-curve: deaths have occurred at a lower than expected rate. Further discussion of this matter can be found in the November 17, 2003 issue of Science (p. 376). The general study of the proposition that "whatever doesn't kill you makes you stronger" is known as hormesis, a word my spell-checker doesn't like because it is a relatively new coinage. But the web may be more forgiving. Some of the best studies of hormesis, particular in regard to radioactivity, have been done by Thormod Henriksen and his statistical group in Oslo, Norway, and his name is worth a Google search.

In one striking and astonishing study, residents of an apartment house in Taipei (Taiwan) were subjected to continuous gamma-radiation for a decade or two, owing to the use of radioactive cobalt-60 in the steel used for construction. [This was not deliberate, and the radioactivity was not discovered for 9 years.] The health of the residents was carefully studied and found to be far superior to that of those who lived elsewhere. In particular, residents had far lower cancer rates. The number of people affected, around

10,000, was large enough that there is no question as to statistical significance.

Because evidence for risk at low levels of any sort of hazard is hard to come by, it is not possible to prove or disprove the LNT hypothesis in most cases. J-curve behavior may be common, or it may be rare. In all likelihood, it will be decades before clinicians and statisticians have collected the data that we'd like to have. In the interim, be wary. Eating broccoli may be good for your health and help prevent cancer. But we don't really know. So don't eat it if you don't like it!

The "noise" in a statistical study is roughly proportional to the square root of the sample size. For example, if we studied the drinking habits of 10,000 randomly selected American men over the course of a year, we see from Figure 6.4 that we can expect about 130 of them to die. The square root of 130 is about 11, so the expected range of deaths is perhaps 110 to 140. Furthermore, the number in any one consumption category will be smaller than 10,000 – let's say 2500 per category, with about 32 ± 6 deaths per category per year. Given that the reduction in death rate is expected to be about 25 deaths in the sample or roughly 6 in each category, our sample size is far too small. The results would be overwhelmed by statistical noise. The minimum sample size for a definitive study in this example is about 100,000. Things only get worse when the risk or reward effects are smaller than those for alcohol – and they usually are. Even larger sample sizes would be needed. It is for this reason that there is still controversy over the effects of low-level radiation doses on the inhabitants of Nagasaki and Hiroshima: Non-lethal exposure of 100,000 people may not be, sadly, large enough to draw definitive conclusions.

When the risk effects are larger, huge sample sizes are unnecessary. In the 50-50-50 rule noted above, we are speaking about a 50% death rate. In a sample size of 100, 50 are expected to perish, or, more correctly, 43 to 57. A very small sample size will tell us most of what we want to know. Statistical noise is a problem only when the effects are small, at "low dose" or "low risk." For this reason, statistical studies on relatively small groups of people are still routinely carried out in testing new drugs and treatments. Such studies may not be able to discover small risks and benefits, but they have the potential to uncover genuinely major effects.

Problems and Exercises

Exercise #1: Compare the alcohol content of ten ounces of beer, six ounces of wine, or one ounce of vodka. How is a "drink" defined by clinicians?

Exercise #2: Studies such as those reported in Figure 6.4 have been criticized because they measure dose by average daily consumption, in effect equating someone who consumes seven drinks once a week ("weekend binger") with someone who takes only one drink a day. Comment.

Exercise #3: Why is it that doctors do not routinely recommend moderate alcohol consumption to their patients who do not drink?

Exercise #4: Studies of the effects of smoking usually divide subjects into two or three groups: Smokers, non-smokers, and (sometimes) ex-smokers. It is generally agreed that a better measure of dose would be "pack-years," in which a pack-a-day for one year counts as one pack-year. Locate a reliable source of information about the risks of smoking and produce a risk-dose graph similar to that shown in Figure 6.4.

Exercise #5: Obesity is generally considered the second-most source of "preventable" death in the United States. Obesity is generally measured by body-mass index (BMI), which is to say weight (in kilograms) divided by height in meters, squared. Thus, a person whose weight is 176 pounds (80 kilograms) and whose height is 5' 10" (70 inches, or 1.78 meters) has BMI = 25.3. Generally speaking, people with a BMI greater than 25 are regarded as overweight; above 35, obese. Locate a reliable source of information about the risks of obesity and produce a risk-dose graph similar to that shown in Figure 6.4.

Exercise #6: In political polling, a sample size of 3000 prospective voters is generally considered adequate to predict the results of an election. What is the expected margin of error in such a poll?

118

Chapter 7: Analysis of Disease Clusters

From time to time epidemiologists find an unusual cluster of health problems, an unexpectedly large number of cases in a small geographical region. They look for a common cause: a chemical spill, an unusual pollutant, a transmissible disease.

What is done is to compare the average or expected incidence of a particular problem with the presumably higher numbers observed in a community or group and ask whether these higher numbers are due to statistical chance -- or not. There is some danger that if too many kinds of diseases are examined simultaneously that completely spurious associations will be found -- a link, to invent a silly example, between water levels of trivalent chromium and automobile accidents -- but we will bypass this danger by focusing on a single disease, childhood acute lymphatic leukemia (ALL).

In the United States, the average incidence of ALL is about 3 cases per year per 100,000 children. The U.S. population was about 280 million in the year 2000, of whom roughly 20% are in the age range 5-13, and the total incidence of ALL was about 1,500 cases per year. Let us divide up the U.S. population of children in two different ways, first, into 56 groups of a million children, and, second, into 560 groups of 100,000, and ask what is the expected statistical distribution of ALL cases in each example. To do so, we will use Poisson statistics, which is generally viewed as the proper way to look for statistically probable clusters. According to the Poisson formula,

$$P(n) = A^n e^{-A}/n! \qquad\qquad n = 0, 1, 2, 3, ...$$

P(n) is the probability of seeing n cases in a population where the average number of cases is A. With 56 groups of a million children, A is $1500/56 = 27$, and the results can be calculated with the assistance of a computer. Our real interest is not in P(n) but in 56 x P(n), the expected number of ALL cases in each of our 56 groups. The results are shown in Figure 7.1, which shows that about 25 cases per group is the most common result. It is not expected that any group will have fewer than 10 or 15 cases, nor more than 45. So a "cluster" of more than 50 cases per million children would be quite unexpected.

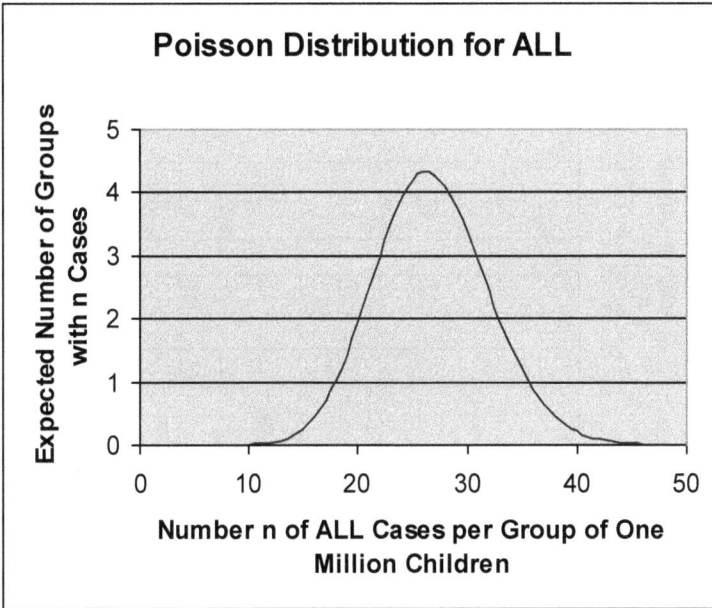

Figure 7.1 Expected Cases of ALL per group of 1 million children

Now we can look at a smaller group size of 100,000 people, where the average number of cases is 2.7 per group. Applying the same formula, Figure 7.2 shows that few groups (38) will have no incidence of ALL in a year, while no group will have more than 9.

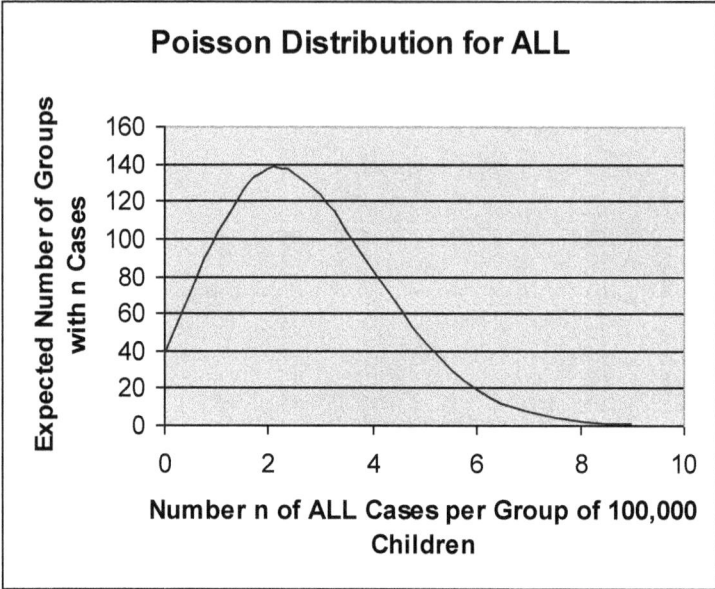

Figure 7.2 Expected Cases of ALL per group of 100,000 children

What are we to understand about Churchill County Nevada, the town of Fallon, where there have been 16 cases of ALL in six years? The population of the county is about 25,000 people, and for the purposes of this analysis, we will take the number of children to be 5,000 so that there are 11,300 groups of the same size in the United States. Taking the expected number of cases per group to be 0.133, Poisson statistics predicts that there will be 9892 groups with no cases of ALL in a year, 1316 groups with one case, 87 groups with two cases, 4 groups with three cases and no group with more than three cases per year. So the cluster is beyond the edge of random statistical probability, and the nine cases diagnosed in the year 2000 stand out as exceedingly improbable due to mere chance. Of course, this analysis provides no clue whatsoever as to a cause; it only says that **some** cause should be searched for. Nuclear radiation is one suspected cause, but as yet the association is far from definitive.

One cause that should always be suspected when there is an improbably large cluster of disease is infectious transmission, which can give rise to a cluster in an obvious way. Viral infection has long been suspected as a cause of leukemia in children -- it is well known as a cause in cats -- but once again the link is uncertain. Incidentally, the rarity of ALL does not in any way rule out an infectious cause. It may simply be that the disease is hard to catch, or that the vast majority of children acquire immunity early in life, or that different ethnic groups have different susceptibilities. One infamous cluster of childhood leukemia occurred in the village of Seascale in England, not far from the Windscale nuclear power station where there had once been (in 1957) a serious accident releasing significant radioactivity to the surrounding region. (The release amounted to a little less than 1% of the release at Chernobyl decades later.) In the period between 1955 and 1984, the rate of childhood leukemia among the children born in Seascale was ten times the average. On the other hand, the rate among children living in Seascale but born elsewhere was zero. This is consistent with two sharply different hypotheses: First, that childhood leukemia is acquired prenatally or in the very first years of life; or, second, that migration of an immune population into a previously isolated population that lacks immunity causes an outbreak, in much the same way that smallpox decimated the indigenous American population after the first European contact. At the moment the second theory seems to me more likely to be right because it correctly predicts leukemia clusters among an isolated indigenous group whenever a large number of outsiders move into the area. However, this is still on the speculative side, and, until the cause or causes of leukemia are known, speculation and statistics are all we have. I wonder, however, if the Fallon Nevada cluster primarily affected native-born children.

The use of Poisson statistics provides a crude start to any investigation of a disease cluster. Statistical methods are commonly used by public health officials to decide whether further investigation is warranted. Quite often, investigations end at this stage: A seeming cluster is found to be no more than chance. This is especially true of cancer clusters: Cancer is quite common, so the probability of finding a larger number of affected people in a small region is quite high. Epidemiologists look for uncommon symptoms, rare diseases that have low occurrences, since the lower the probability, the more likely an apparent cluster has a genuine common origin. Then the hard work begins, trying to find more specific commonality than simple geographical location. This effort requires labor, skill, and judgment, and it is enormously useful if there have been prior epidemiological studies.

Consider, for example, a family whose members all come down with headaches of varying degrees of severity at much the same time. Headaches are very common, and a viral infection is a strong candidate for a common origin. But every physician knows there are some routine questions to be asked that center on the possibility of carbon monoxide poisoning: Who stays at home? Who rides in a car? What method is used to heat the house? The answers to these questions, plus an examination of the severity of symptoms of each family member, can quickly rule in or out the role of carbon monoxide, and a blood-work follow up of the most severely affected family members can settle the matter quickly. But prompt diagnosis depends on the fact that carbon monoxide poisoning is well understood, and knowledge of its effects is widely disseminated. A cluster of headaches, however common, warrants immediate investigation, since carbon monoxide poisoning can be fatal, and since diagnosis is quick and accurate.

The relationship between cause and effect is not usually so well understood when the disease is uncommon. We humans live diverse lives, and it is impossible to devise medical questionnaires that cover all of our activities. The connection between smoking and lung cancer took decades to document despite the fact that the health effects of smoking are – as we now know – huge. When it comes to health, statistical studies are slow, expensive, and usually misleading, but absolutely essential to identifying real hazards. The inefficiency of such studies does not mean epidemiologists should not keep on trying. The cause of AIDS was discovered through statistics, following the usual progression of anecdotal observations (Karposi sarcoma was found only in homosexual men) to more formal statistical analysis (homosexual activity correlates with an increased rate of mortality) to careful biological analysis (a retrovirus is the causal agent). Finding such connections is hard work, and preliminary studies are usually inconclusive or wrong. It isn't easy, in the face of conflicting evidence, to maintain a balance between credulity and skepticism. What is required is a suspension of final judgment.

The latest hazard to cause widespread concern is mad cow disease, or, as it is known in humans, "new variant" Creutzfeldt-Jakob disease (vCJD), which, unlike classic CJD, affects young people[14]. Both forms of the disease are rare: the death rate due to conventional CJD is one in a million per year, while vCJD has probably killed fewer than 200 people worldwide. How was the cause of such a tiny incidence of disease determined? First, vCJD appeared as a cluster of sorts: 95% of the worldwide cases involved people who lived in the United Kingdom. Second, vCJD is easy to

---

[14] The median age of a CJD patient at death is 68 years, compared to 28 years for patients with vCJD.

diagnose: The symptoms are nearly unmistakable, and can be confirmed by autopsy: Victims' brains resemble those of cows that have died from bovine spongiform encephalopathy (BSE). Third, a form of vCJD (Kuru) was known to occur in the indigenous Fore tribe of New Guinea, which once practiced cannibalism. There was also more direct evidence: BSE can be transmitted from one cow to another by injecting the brain of a healthy cow with an extract from the brain of a sick cow, leaving no doubt that BSE is an infectious disease, although, strangely enough, routine bacteriological tests show that the infectious agent is neither a bacterium nor a virus. There is still a great deal about vCJD and BSE that is not well understood: If the agent is a protein fragment or prion, why isn't it digested in the stomach, how does it cross the blood-brain barrier, why is vCJD so rare, what is the incubation period, and why is eating BSE-infected **meat** a hazard when BSE affects **brain** tissue? Fortunately, we do not need immediate answers to these questions in order to take appropriate action, which is to test "downer" cows for BSE and ban cattle feed that contains animal meat and byproducts. New variant CJD is likely to become vanishingly rare.

Statistical analysis of a disease cluster does not usually tell us the cause. The next step in an analysis is to identify what the patients have in common. Sometimes this is relatively easy: lung cancer, cigarettes; Karposi sarcoma, male homosexual sex. More often a connection is hard to find, as is the case with childhood ALL. The connection is often obscured because the disease has a long incubation period, as is the case with AIDS, lung cancer and vCJD. Finding a cause – or, speaking in legal terms, a probable cause – is not as easy as many people seem to think.

Problems and Exercises

Exercise #1: A cluster of CJD cases was traced to a restaurant at a racetrack in New Jersey, although the Centers for Disease Control (CDC) says there is no connection. Read the latest reports on this subject. Do you agree with the CDC?

Exercise #2: In analyzing the incidence of ALL, we looked at the population of children rather than the total population. Analyze the expected frequency of ALL using the entire United States population as a basis. Does this make the nine cases of ALL in Fallon County diagnosed in 2000 seem more or less likely to be due to coincidence? Which approach is better?

Exercise #3: Read "A Civil Action" by Jonathan Harr, as well as commentaries on the case that are posted on the web. How might Poisson statistics be used in analyzing data in this case? What population should be used in the analysis, the total population of Woburn, Massachusetts or the population that used water from wells G and H?

Exercise #4: For a wonderful report on how the U.S. Centers for Disease Control (CDC) investigates disease clusters, read "The Medical Detectives" by Berton Roueche.

Exercise #5: A small company receives 520 orders a year for a particular product, that is, ten orders per week on average. In the course of a year, what are the maximum and minimum numbers of orders they can expect? What is the maximum and minimum expected in the course of a day, assuming a five-day week?

Chapter 8: Autism and Mercury

Statisticians usually resort to the linear dose hypothesis because there is rarely any useful data for low-dose exposures. This is especially true when a section of the population is exposed to environmental pollutants, such as fallout from atomic testing, particulates from smokestacks or chemicals from a toxic waste dump. Trace metals, such as lead, mercury, and chromium are a particularly well-studied concern, in large part because the concentrations of these metals can be measured with a very high degree of accuracy. However, even these simple metals present some difficulties, because they occur in different forms. Lead, in the soluble form found in paints or in the organic form once used as a gasoline additive, is a hazard; lead sulfide, found in minerals, in harmless. Chromium metal and hexavalent chromium are hazardous; trivalent chromium is entirely safe. For that matter, chlorine gas is a poison once used in the First World War, while chloride is a major component of seawater and all plants and animals. Mercury is present in several different forms, including the liquid metal found in thermometers, the amalgamated form used in dental fillings, the inorganic form found in calomel, and the organic forms such as methyl mercury and ethyl mercury; each form has a different level of danger. Here we will focus on the organic forms of mercury because they are exceptionally well studied and are much in the recent news: Exposure to organic mercury has been linked to a rise in autism.

The earliest studies on organic mercury were concerned with poisoning associated with the consumption of contaminated fish in Japan. Deaths were associated with eating fish that contained 5 to 10 parts per million mercury (ppm Hg). Other studies examined people poisoned by accidental exposures to mercury-containing pesticides, and to people who ate the meat of animals that ate food contaminated by pesticides. It was quickly established that a blood level of 1 ppm Hg in humans was sufficient to lead to severe neurological damage (loss of hearing or sight); levels much above 1 ppm were usually fatal. This led to a recommendation by the health authorities that fish with a concentration of mercury above 0.5 ppm Hg should not be eaten, and consumption of fish with lower levels of mercury should be limited.

If a person weighing 50 kilograms (110 pounds) eats a quarter kilo of fish (half a pound) containing 0.5 ppm Hg, the concentration of mercury in the body should increase by $(0.25/50 \times 0.5) = 0.0025$ ppm, assuming all the mercury in the fish is absorbed. This level appears to be safe, although it is only a factor of 400 below the level associated with neurological damage. One important question is the rate at which humans excrete mercury; 0.0025 ppm Hg may be safe, but what happens if one consumes fish daily or weekly? The answer is not known.

Studies of "uncontaminated" people show that mercury levels in blood are usually in the range 0 - 2 micrograms per deciliter of blood, that is, up to 2×10^{-6} grams Hg per 100 grams of blood, or 0.02 ppm. This suggests that a whole-body level of mercury of 0.02 ppm is safe, while levels around 1 ppm are extremely dangerous. At first glance, this is an exceptionally narrow range, hard to understand unless mercury is quickly excreted or converted by cells to a harmless form. What we need is more information on groups of people exposed to low doses of mercury.

As it happens, the experiment has been done, unfortunately or otherwise. Millions of children have been exposed to mercury through vaccinations. Until recently, a preservative containing organic mercury was routinely used in vaccines. Each vaccination contained less than 0.01% mercury, but children are given so many shots that the total exposure to mercury in early childhood routinely reached over 100 micrograms. Assuming for simplicity that a child weighs 10 kilos, this is roughly a whole body level of 0.01 ppm, safe enough for adults, presumably, but potentially dangerous, as a neurological toxin, to developing children. This level has been associated with an alarming rise in the rate of autism, which now affects about one in every 100 children.

What does the linear dose hypothesis have to say in this case? If we extrapolate the danger level in adults downward, we might expect that one in one hundred children would be affected with some sort of neurological damage, so the actual rate of autism appears consistent with this assumption, perhaps a bit low, given that children ought to be more susceptible to neurological damage, since their brains are still developing, and given that autism is only one of many forms of potential damage.

The medical authorities have done a far more detailed and sophisticated analysis of the rate of autism than the very rough examination presented here. Rates of exposure vary widely, depending on whether a child received a vaccine containing mercury or some other preservative. Not all children are vaccinated. The detailed studies show no "statistically significant" connection between mercury and autism. Nevertheless, much doubt remains, particularly among the parents of the affected children, and their lawyers.

I invite the reader to come to his or her own conclusion, and, having reached that conclusion, to search the web for information on thimerosal, the mercury-containing preservative once widely used in vaccines. This preservative has been eliminated from children's vaccines. Since autism typically can be diagnosed before the age of two, we will know within a very short time whether mercury is a significant contributor. Epidemiological and medical studies rarely end with the prospect of a conclusion so clearly in sight.

Post Script: By 2010 the conclusion was clear: Mercury poisoning from vaccines has nothing to do with autism.

Problems and Exercises

Exercise #1: Assume there are approximately 675 million cells per cubic centimeter of body tissue (see Chapter 5). About how many cells are in a person weighing 50 kilograms? If this person has a whole-body concentration of 0.02 ppm Hg, what is the total body weight of mercury? What is the number of moles of mercury in the body (the atomic weight of mercury is 200 grams per mole)? How many atoms is this? On average, how many atoms of mercury are found in each cell?

Exercise #2: The gap between a clearly dangerous blood level of mercury (one part per million) and a clearly safe level (0.02 ppm, the concentration found in presumably uncontaminated people) is relatively narrow, only a factor of 50. Is this consistent with the linear dose no-threshold (LNT) hypothesis discussed in Chapter 6?

Exercise #3: In December of 1990 the CBS television program "60 Minutes" ran a segment exploring the dangers of mercury amalgam used to repair decayed teeth. Follow up on this story. What is the current thinking on the use of mercury amalgam?

Part III: Energy and the Future

Chapter 9: Energy Storage

In Chapter 5 I mentioned that electric power is one of the true miracles of modern technology: Energy is delivered, on demand, over huge distances, right to your home or office whenever you want it for as long as you need it. Electric power can be generated in a great variety of ways, from coal or nuclear, oil or gas, hydro or wind or solar. Because your demand is shared with millions of other people, energy storage isn't essential, and when total demand varies throughout the day and the seasons, power plants can be brought online or banked down to meet the changing needs, so that the system nearly always operates at high efficiency. The grid, like the net, connects us to our neighbors, allowing us to share the costs of expensive capital facilities. Furthermore, supplies of coal (and nuclear fuel) are abundant, so electrical shortages are not likely to be part of our future unless, for environmental reasons, we decide to shut down our coal- and nuclear-fired power plants.

My summer home is on an island that is not connected to the grid. We generate our own electricity, in earlier days with gasoline-powered generators, and nowadays with solar. The sun doesn't shine at night or on cloudy days, so we store the energy in batteries and hope that it isn't cloudy for more than a few days in a row. We keep the generators for backup and for the vacuum cleaner. Our system can handle small power tools and lights and radios and even a high-efficiency refrigerator, but we don't have toasters or hair dryers or a microwave oven. Off-grid uses like ours are uncommon, but there is a sufficiently large market in homes and lighthouses and offshore platforms to support a growing solar power industry.

The largest off-grid market is transportation: Cars, trucks, planes, and ships. In each case, fossil fuels are still the preferred medium of energy storage. Unfortunately, technology has not found a way to make fossil fuels from electricity, so we take what we need from deep underground and refine it into different products: Kerosene, gasoline, or diesel fuel. As energy storage media, these fuels are spectacular. A liter of gasoline weighs only 0.74 kilograms (compared to one kilogram for water) but releases 36 million joules when burned in air. That's about 10 kilowatt-hours! Of course, only a small fraction of this energy, about a fifth, can be turned into useful work, but that's still 2 kilowatt-hours in a liter. There is an environmental cost, of course, 2.3 kilograms of carbon dioxide (CO_2) released to the atmosphere per liter of gas, not to mention pollution due to nitrogen oxides and unburned hydrocarbons. But most energy consumption has an environmental cost.

What are the alternatives for energy storage? What we are really looking for is a way to store electrical energy, although using electricity does not necessarily solve all the environmental problems. The possibilities include batteries, supercapacitors, fuel cells, and hydrogen, not to mention flywheels and pumped hydroelectric. Each one has its own set of limitations.

Here are some definitions, units and conversion factors you will need to understand the material presented in this chapter. Some of this information was presented in Chapter 0. **1.** A joule is a unit of energy, while a watt is a unit of power. A watt delivered for one second – a watt-second – is equal to a joule, or, to put it the other way around, a watt is a joule of energy per second. A watt is also the power delivered by an ampere of current at one volt. A watt-hour is 60 x 60 (3600) joules. **2.** A coulomb is a unit of electrical charge equal to an ampere of electrical current delivered for one second. An ampere-hour is 3600 coulombs. **3.** A faraday is a mole (6.02×10^{23}) of electrons. It is equal to 96,500 coulombs. Thus 6.94 grams of lithium (a mole), which delivers a mole of electrons when ionized, produces 96,500 coulombs, or 96,500 ampere-seconds, or 26.8 ampere-hours. At three volts, this is about 80 watt-hours. **4.** A farad (not to be confused with a faraday!) is a unit of electrical capacity. Its units are coulombs per volt. The energy that can be stored in a capacitor is ½ CV^2, where C is the capacitance in farads and V the maximum voltage.

Batteries are a good place to start. The familiar car battery, used for starting (and lights, radio, etc.) weighs about 30 kilograms and stores about a one-kilowatt hour of useful energy. (A car battery is rated at 12 volts, and can deliver about a hundred ampere-hours if discharged slowly. That's 1200 watt-hours.) So two car batteries can deliver about the same energy as a liter of gasoline while weighing 80 times as much. Ouch!

Lead-acid batteries, the type used in cars, are a pretty good choice for stationary applications, such as my island house, where weight is unimportant. Their main drawback is that they age, and the more they are used, the faster they age. Lead-acid batteries die quickly if they are repeatedly discharged by more than 50% of capacity, a circumstance that is hard to avoid when the sun doesn't shine for days on end. Discharging a lead-acid battery by more than 80% (80% depth-of-discharge or 80% DOD) will ruin it even more quickly. In solar applications, it is hard to keep a lead-acid battery in service for more than five years.

There are better batteries, both from the point of view of energy density (stored energy per weight) or cycle life. Nickel-cadmium (NiCad), nickel-metal hydride (NiMH) and lithium ion (LiON) are all better than lead-acid. But nothing comes close to lead-acid in terms of cost. Not that lead-acid batteries are cheap: A car battery only stores about a dime's worth of electricity, although this isn't quite a fair measure since the dime can be put in and taken out repeatedly. But you'd need 1000 cycles from the battery at 50% DOD to amortize the $100 cost even if you didn't have to pay anything to recharge it.

What are the prospects for better batteries? In terms of energy density, lithium-ion batteries are close to the theoretical limit, at least in principle. The chemical reaction is $Li^+ + e^- \Rightarrow Li$, where e^- represents an electron. The voltage for this reaction is about 3.5V, and the atomic weight of lithium is only 6.94, from which it can be shown that a kilogram of lithium can store nearly 14 kilowatt hours. That's even better than gasoline! If only it were possible in practice. Lithium atoms (Li) generally require a storage medium, typically graphite, six atoms of carbon[15] per lithium atom. That bumps up the molecular weight from 6.94 to 6 x 12 +6.94, or 78.9. We just lost a factor of eleven! Positive lithium ions (Li^+) also require a medium, typically manganese dioxide, MnO_2, two molecules per lithium atom. That adds another 174 to the molecular weight, for a total of 253. Now we're down by a factor of 36. And of course the packaging materials add still more weight. In practice, the energy density of LiON batteries is about 0.15 kilowatt-hours per kilogram, down by a factor of 100 from the theoretical, and down by a factor of 18 from gasoline.

---

[15] The atomic weight of carbon is 12 grams per mole.

This is not as terrible as it sounds. A car needs perhaps 100 kilowatt-hours of storage to go 300 miles between refuelings. Lithium batteries could provide this much energy in a 600-kilogram package. That's comparable to the weight of a car engine. So a battery-powered car is technically feasible provided that LiON batteries can withstand thousands of repeated charge and discharge cycles. This appears to be a real possibility. Cost, however, is another matter. Safety is another potential problem.

I think two conclusions can be drawn from the information just presented. First, battery-powered cars are technically feasible. Just look at Tesla. There is already a real (unsubsidized) market in vehicles that travel only a few miles a day, such as postal vans and forklifts. Second, there is ample room for technical improvement and cost reduction, although I caution the reader that progress in battery development has always been painfully slow despite what seem to me to have been ample levels of funds. The market for good batteries is not small; the space program has poured money into research. (Communication satellites also need batteries.) Of course, some consumers are willing to pay for all-electric vehicles, regardless of cost, and this consumer demand is rapidly driving down the price of electric vehicle (EV) batteries, as shown in Figure 9.1. The price is currently not far from what experts view as "the magic price", $100 per kilowatt-hour, basically the price of a lead-acid battery but with far less weight and far more cycle life.

Price of Electric Vehicle Batteries (Lithium)

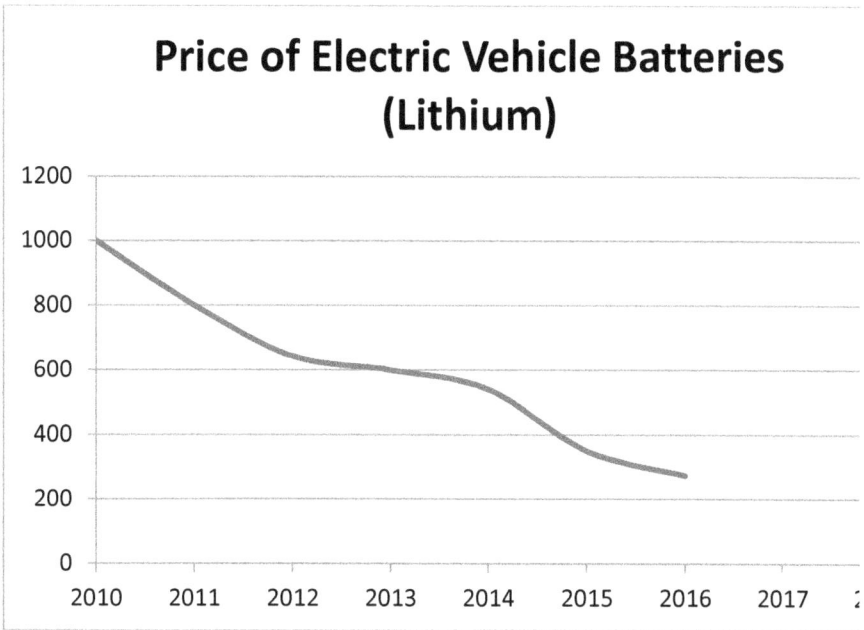

Figure 9.1 The Declining Cost of Lithium EV Batteries ($/kWhr)

https://cleantechnica.com/2017/12/11/batteries-keep-getting-cheaper/

What about the other major possibilities for energy storage, ultracapacitors, fuel cells, and hydrogen? Unlike a battery, a capacitor stores electrons in a thin film of conductive material, rather than by converting electrical energy to chemical energy. Typical capacitors, the kind found in all electronic devices (radios, television sets, cell phones) store only a minuscule amount of energy. Capacitance is usually measured in microfarads, and a big old-fashioned TV capacitor might reach 10,000 microfarads or 0.01 farads. At (say) 100 volts, this is only 50 joules of energy. It would take 72,000 such capacitors to store a kilowatt-hour. This is not encouraging. A modern "ultracapacitor" the size of a D-cell battery can store about 1000 joules, but that figure still implies 3,600 cells per kilowatt-hour. (By comparison, a D-cell nickel-metal hydride battery can store about 6 ampere-hours or roughly 28000 joules.)

Traditional capacitors can be charged and discharged endlessly, without degradation. Furthermore, they can be charged or discharged at very high rates, which means that "refueling" times can be very short. This is their attractive side. The negatives are cost, weight, and volume.

Nevertheless, new types of capacitors have been developed with much higher energy densities, evidently solving the weight and volume problems while preserving the promise of very high rates of charge and discharge. The questions that remain pertain to cycle life and cost. It is too early to know whether these difficulties can be overcome, but prototype cars have been built using ultracapacitors. As with cars powered by lithium batteries, they are sold primarily on their ability to accelerate like a rocket: They can deliver their stored energy over very short periods of time. But it is too soon to evaluate their commercial potential. It does seem likely that a mixed system of batteries and ultracapacitors can provide longer-term energy storage (batteries) with short-term high-power discharge rates (ultracapacitors) in a way that is optimal for both rapid acceleration and highway cruising.

As for fuel cells, they are really not much different from batteries. In a battery, the chemical materials that store energy are usually solids that can be ionized. The ions move only in response to the flow of electricity. In a fuel cell, the reactive chemicals are liquids or gases, which can be moved into position as required. In principle, this reduces the areas and volumes required to place the chemicals near a reactive surface. This is an attractive idea. Fuel cells using methanol and air (oxygen) are nearing commercial feasibility. While attractive for small applications, such as laptop computers and cell phones, their benefit in transportation applications is limited, since they still produce carbon dioxide, and the fuels cannot be generated from electricity.

Fuel cells based on hydrogen (H_2) seem, at first glance, to be much more attractive. The energy that is released when hydrogen reacts with oxygen in the air is comparable, on a weight basis, to the energy stored in lithium; hydrogen can be generated, quite easily, from electricity; when consumed in a fuel cell, the only product is water; and fuel cells using hydrogen have been used in spacecraft for many decades. There is absolutely no question as to technical feasibility. Why then are so many scientists skeptical about the future of hydrogen as a transportation fuel, as an energy storage medium? The prime answer, quite simply, is that hydrogen is not cheaply and readily compressed. It is not the weight per unit of energy storage that is worrisome, but the volume. Hydrogen cannot be liquefied unless it is chilled to extremely low temperatures, and refrigeration is expensive and inefficient. Hydrogen can be compressed without liquefaction, but the steel tanks needed are themselves heavy or bulky, so much so that the perceived advantages of hydrogen on the basis of weight are lost. There is a third way to store hydrogen, via reversible reaction with certain metal powders to produce a metal hydride, in the same way that hydrogen is stored in nickel-metal hydride batteries. But the battery market for laptops and cell phones has shown that the energy storage density in lithium-ion batteries is better than in nickel-metal hydrides, which are already yesterday's technology. True, the hydrogen storage problem may be solved by newer techniques not yet ready for commercialization. But the storage problem has been around for decades and progress has been almost non-existent. And there are other problems with respect to hydrogen (generation, storage, transportation, safety) that make me think that lithium batteries are a better bet. Ultracapacitors have a serious chance, but for the time being, I'll put my money on lithium. Or maybe, just maybe, an old but improving technology, zinc and air (oxygen). (See Exercise #10),

It would be wonderful if we could make liquid fuels from electricity. That's another bet I refuse to take. The difficulties are way beyond hard. But there are alternatives.

148

The starch and sugars in plants can be converted to alcohol, and nature makes liquid fuels directly, in the form of vegetable oils: Palm oil, corn oil, safflower oil, olive oil, peanut oil, rapeseed oil, and so on. All of these oils can be converted, quite easily, to diesel fuel, or used (with a little heating) as is. Worldwide production is only a few percent of what is needed to replace petroleum fuels, but in principle, we could simply plant more palm trees. Vegetable oil at your local supermarket is only about three times[16] as expensive as diesel fuel (five times as expensive if we subtract the fuel taxes). Biodiesel fuels hold real promise as an energy storage medium, and their use releases no more carbon dioxide to the atmosphere than the plants removed. We don't even need electricity: Sunlight[17] is the energy source.

Does this sound too good to be true? Perhaps it is. The imbalance between the demand for transportation fuels and the supply of vegetable oils is huge. Biotechnology may offer an alternative. Perhaps oils can be produced by bacteria or other simple organisms. But increased production is not the only way to solve the transportation fuel problem. Reducing demand by increasing efficiency of use is a very

---

[16] Used cooking oil from the fast food industry is much cheaper, and current prices for biodiesel fuel are comparable to prices for fossil diesel.
[17] Sunlight isn't the only energy source: Fossil fuels are consumed in making fertilizer, and there is some question as to whether the fossil energy consumed in making vegetable oil is comparable to the energy stored in the oil. This is a hard question to answer, but it is clear that growers would reduce fertilizer energy inputs if fossil fuels were more expensive.

reasonable alternative. Volkswagen has produced an experimental car that gets 200 miles to the gallon; hybrid cars and turbo diesels on the market today get over 50. China has announced fuel efficiency standards that far exceed the U.S. standards. I don't think the world will produce cars that routinely exceed 200 mpg in my lifetime, but if that happened, the world could run on biodiesel. There is, I think, ample reason to be hopeful that the transportation fuel problem will be solved before the world "runs out of oil," by which I mean the time when fossil oil becomes more expensive than vegetable oil.

Conservation, by which I mean improved efficiency, is typically expensive in the short run and cheap in the long. Hybrid cars cost more in the short run, and someday perhaps less over their lifetime: Capital costs are higher, annual costs are lower. It's a tradeoff that is hard for a consumer to make. Government regulation can help; so can common sense. My summer house in Maine uses about two or three kilowatt-hours of electricity per day, while my winter house in Massachusetts uses thirty. Why do we use so little in Maine? Because electricity is expensive, a dollar a kilowatt-hour (I estimate) compared to a dime in Massachusetts. So we use compact fluorescent and LED lights in Maine, which have three to five times the efficiency of the incandescents we used in the past. Because we dry our wet clothes on a sunlit line, not in a dryer. Because our refrigerator in Maine is three times as efficient as the one in Massachusetts. Because we turn off the lights in Maine when we leave a room. Because lugging gasoline to a generator is a time-consuming nuisance. Solar power makes great sense when you're off the grid, but perhaps only when you're off the grid. As we shall show in the next chapter, off-grid uses may be the only places where solar makes complete sense, at least at present prices. The technical and economic inadequacies of energy storage are still severe, but prices are changing rapidly. In the very long term, it might be cheaper to electrify our roads and streets, as we do with trolley cars, than to try to carry the stored energy with us in our cars and trucks. Maybe only airplanes and ships will run on fossil fuels. And maybe the price of lithium batteries will become so low that they can be used in conjunction with wind and solar to provide most of the world's energy needs. We'll see. There are alternatives, choices to be made, and with luck and intelligence and time we'll choose well.

Problems and Exercises

Exercise #1: Find out what is meant by the "self-discharge rate" of a battery. Compare self-discharge rates for the four principal rechargeable batteries on the market today, i.e., lead-acid, nickel-cadmium, nickel-metal-hydride, and LiON. For comparison, what is the self-discharge rate of an ordinary alkaline (non-rechargeable) flashlight battery?

Exercise #2: If a car gets 30 miles to the gallon, which it converts to useful work at 20% efficiency, what energy (in kilowatt-hours) is required to drive 300 miles?

Exercise #3: The "cycle life" of a battery is generally defined as the number of 100% DOD cycles the battery can undergo before losing 20% of its capacity. (For some batteries that cannot withstand a full discharge, it is defined as half the number of 50% DOD cycles.) Compare cycle lives of the four types of batteries mentioned above.

Exercise #4: The "C-rate" of a battery is defined as the time required to charge or discharge it without damage. For example, if a 200 ampere-hour battery can be charged safely at 10 amperes, this a 20-hour rate, often expressed as "C/20", capacity divided by 20 hours. Compare manufacturers' recommended maximum C-rates for the batteries mentioned above for both charge and discharge.

Exercise #5: Compare the oil production rates for different types of plants. Which plant produces the most oil per acre? How does this compare to the production rate for corn oil?

Exercise #6: Compare the energy density of a commercially available methanol fuel cell with a LiON battery.

Exercise #7: The Toyota Prius, a hybrid car, uses a small (50 kg) battery pack consisting of about 168 1.2-volt nickel-metal hydride cells. The maximum power output of the battery pack is about 27 kilowatts (36 horsepower). If the "energy density" of the battery is about 60 watt-hours per kilogram, what is the discharge C-rate at peak power output?

Exercise #8: Compare the wholesale (commodity) prices of vegetable oil and crude oil.

Exercise #9: There are many different chemistries in lithium-ion batteries, each with distinctive and different properties. As an example, compare the uses of cobalt oxide with iron phosphate as to safety, energy density, and power density.

Exercise #10: Zinc metal is widely used in hearing-aid batteries. But zinc, when exposed to air, slowly decomposes to zinc oxide, even while not producing electricity, which is why hearing-aid batteries are protected until use by a glue seal. How do manufacturers propose to overcome this problem of self-discharge? How do they propose to regenerate zinc metal from zinc oxide?

Chapter 10: Solar Economics

The costs of renewable energy sources such as solar, wind and hydro are primarily capital costs: Pay once and get years of benefits. Annual costs, for maintenance and repair, are usually relatively low. It is difficult to compare the cost of capital with the cost of fuel, and the economic analyses can get quite complex, involving present and future interest rates, inflation rates, fuel costs, and so on. Purchasing a renewable energy system is an investment, and like all investments, it can turn out well or badly. As stock market advisors are forever saying, past performance is no guarantee of future success (and past failure is no guarantee either!). Because of these uncertainties, it is sensible just to look at the raw numbers in an unsophisticated way without trying to decide right away whether or not renewable energy is presently economically feasible.

If I were installing an off-grid solar electric system today, I'd probably start with two 100-watt modules at a total cost of just over $300. For storage, I'd use a 12-volt deep cycle battery with a capacity of 390 ampere-hours costing about $600. Then I need a charge controller to prevent damage to the battery, an inverter to convert 12 VDC to 110 VAC, and some wires. These items add $500 to $600 to the cost, depending on quality. I'd do the installation myself, so I won't include that cost. In any case, the work is easy, and the solar panels only weigh 25 pounds apiece. So, for under $1,500, I'll have a small solar electric system. What will it provide? The solar collectors are each rated at 6.2 amperes, 16.1 volts in direct noon sunlight. At 14 volts, which is needed to charge the battery, I might get 12 amperes total on June 21 at noon, when the sun is highest in the sky. Neither June nor noon last long, and over the course of a typical bright summer's day I might get five hours of good performance. That's 60 ampere-hours, so the battery I selected can store the output of six good days. Or three days if I don't discharge it by more than 50%. I'm hoping not to have more than two or three overcast days in a row, although even on cloudy days the panels will collect a little energy.

This system could collect more energy if the solar collectors were steered so as to constantly face the sun, but solar tracking systems are expensive, and it is not clear that the benefits outweigh the cost. I'll change the tilt angle of the collectors every month, so even in spring and early fall I'll get a full 60 ampere-hours on a bright day. Not every day is cloudless, and I'll be lucky to average 30 ampere-hours a day over the course of the summer.

There are other inefficiencies. Energy goes into the battery at 14 volts and comes back out at 12.5. If there are many sunny days in a row, the storage batteries will be filled to capacity and the controller will shut off the collectors. All in all, the system will probably produce 350 watt-hours – not quite half

156

a kilowatt-hour – each day. These assumptions may seem a bit pessimistic, but it is well to remember that under the most optimistic set of assumptions (no clouds; no inefficiencies) the limit would be around one kilowatt-hour per day. Maybe I should have spent an extra $150 on another solar panel!

Over the course of a summer, the system might produce 50 kWhr. Over the course of a year, 100 kWhr. Not much, but it's paid for! How long will the system last? The solar panels are guaranteed for twenty years (with no more than 20% degradation in output), so I'll ignore the degradation and cost out the electricity over a 20-year period. I'll have to buy new batteries every so often, adding maybe $1200 to the lifetime cost, so the final 20-year output is 2,000 kWhr (year round) or 1,000 kWhr (summers only) at a total cost of say $2,700. That's over a dollar a kilowatt-hour; $2.70/kWhr for summers. You can tweak the assumptions all you want, even move the system to Arizona or to the equator, but you won't get the cost down to a dime (or twelve cents), which is what I'd pay if I were attached to the grid.

The costs of solar electricity can be reduced substantially by eliminating the storage component with a direct tie-in to the grid – at least in principle. At present, tie-ins are nearly as expensive as storage. At least all the electricity generated by the panels will be used; there is no question of exceeding the battery storage capacity. In a very favorable climate my two panels might generate up to a kilowatt-hour a day, 360 kWhr a year, 7,200 kWhr over the course of 20 years. Ignoring the cost of the tie-in (we are shooting here for total optimism), this could bring the cost down to 14 cents a kilowatt-hour, which is not quite cheap enough to be competitive, but close enough to keep in mind.

Environmentalists have long argued that the price we pay for electricity does not reflect the real costs. There are "exogenous" costs associated with coal mining, air pollution and waste disposal that are not reflected in the price. While this is surely true, it is hard to believe that the "true" price is double what we are paying now.

There is another economic question that should be raised: What about the energy consumed in manufacturing a solar collector? Are we sure that the energy produced over the lifetime of a solar panel will exceed the energy used in making it? We can make a crude estimate in the following way. In the year 2000, U.S. energy consumption was 100 quadrillion BTUs (100×10^{15}), while the gross domestic product was 9.7 trillion dollars (9.7×10^{12}). This is 10.3×10^3 BTUs per dollar of product, or 10.9×10^6 joules. A joule is a watt-second, so this is about 3 kilowatt-hours per dollar of product or 3,000 kilowatt-hours for my two solar panels. (I'm ignoring, for simplicity, the costs of the batteries and other components). This calculation is a bit of a cheat, because the energy that goes into manufacturing is primarily heat, not electricity, and the value of electricity is much greater than the value of heat. As electricity, the cost of my panels might only be 1,000 kWhr. So, over their useful lifetimes, my panels will probably generate more energy than went into making them.

The example outlined above is, as promised, unsophisticated. It does point the way toward a true economic analysis, but at current price levels, this seems pointless. Solar energy cannot compete at the moment with electricity from the grid except in places that are exceptionally sunny. This does not mean solar energy is infeasible. There are many off-grid applications (including communications satellites!) in which paying over a dollar a kilowatt-hour is cheaper than the alternatives. These applications drive the competitive research and development cycle that drives down costs and produces cheaper products. Solar electric power generation has a growing place in the energy market.

Wind energy[18] looks much better than solar, although I cannot recommend wind turbines for small domestic applications. There are economies of scale that make large turbines much cheaper per kilowatt than backyard windmills. Capital costs for wind turbines are now down to about $1,600 per kilowatt, comparable to the cost of the solar panels I mentioned above. As with solar, much of the

---

[18] See "Wind Energy Explained" by J.F. Manwell, J.G. McGowan and A.L. Rogers, John Wiley & Sons, Ltd., New York (2002).

economic analysis depends on the renewable source – how many hours a day the skies are clear, how many hours a day the wind blows at a useful speed. It appears that wind energy can get well below the dime per kilowatt-hour goal. As with solar, market penetration has been rapid, although there are concerns about the presumed unsightliness of large wind turbines. This seems to be largely a problem of unfamiliarity: In Denmark and the Netherlands, where wind energy has always been popular, most people who live near windmills find them attractive.

Large wind turbines are usually connected directly to the power grid, so there is no storage problem (as long as wind contributes only a small percentage of overall electrical production). Furthermore, grid-connected wind energy is a relatively young technology. The present emphasis is on rugged systems that can resist hurricane-force winds and that have low maintenance costs; past efforts to produce lighter and less-expensive systems have generally ended in failure. However, now that wind farms are proving successful, a new generation of less expensive designs can be envisioned, and it seems likely that wind energy will play an increasing role in worldwide electrical supply. The cheapest source of renewable electrical energy production is of course hydropower, which presently supplies about 10% of U.S. electrical demand and a far higher percentage in mountainous and thinly populated countries such as Switzerland and New Zealand. However, hydropower is not a growth industry: much of the world's hydro resources have already been exploited. In a less populous or more efficient world, hydroelectric power would be the energy source of choice, at least in mountainous regions where the environmental impact seems relatively small. In less vertiginous areas, "low head" hydro, often produced by damming rivers and streams, has a much more significant impact by cutting off fish spawning grounds. It is often the case that the economic value of the hydroelectricity produced at these dams is less than the economic loss due to reduced catches of freshwater fish.

Problems and Exercises

Exercise #1: Compare prices for electric power in various parts of the United States. Where is electricity the cheapest? Where is it the most expensive? What accounts for the differences?

Exercise #2: Environmentalists argue that the prices we pay for electricity do not reflect the real costs, which include detrimental health effects associated with coal mining and burning, global warming, etc. They have suggested a "carbon tax" on coal, gas, and oil that would raise the price of fossil fuels and encourage both energy conservation and the use of renewable sources. What range of carbon tax rates has been suggested?

Exercise #3: The unsophisticated economic analysis in this chapter ignores many important factors, such as the expected rate of return on invested money, inflation, tax rates, and so on. Perhaps the most important missing factor is the expected rate of increase in electricity prices, over and above inflation, in the future. A capital investment that hedges against rising prices may make sense in the long run. However, future prices are unknowable; what is usually done is to look at the trend in past prices. Compare the average rate of increase in the cost of electricity in the United States with the rate of inflation over the past ten years.

Exercise #4: In the United States, wind energy farms currently receive a tax subsidy. How much does this subsidy amount to in terms of price per kilowatt-hour?

Chapter 11: Energy and Population

When I was born, the population of the United States, and for that matter, the population of the world, was less than half what it is today. Both U.S. and world populations have doubled in the past fifty years. Americans are used to this sort of rate of growth: The U.S. population has been doubling in far shorter time intervals since Europeans first settled the country. For the rest of the world, such rapid growth is a much more recent event. As shown in Figures 11.1 and 11.2, the world growth rate peaked in 1965, while the U.S. growth rate has been declining (somewhat erratically) since the civil war.

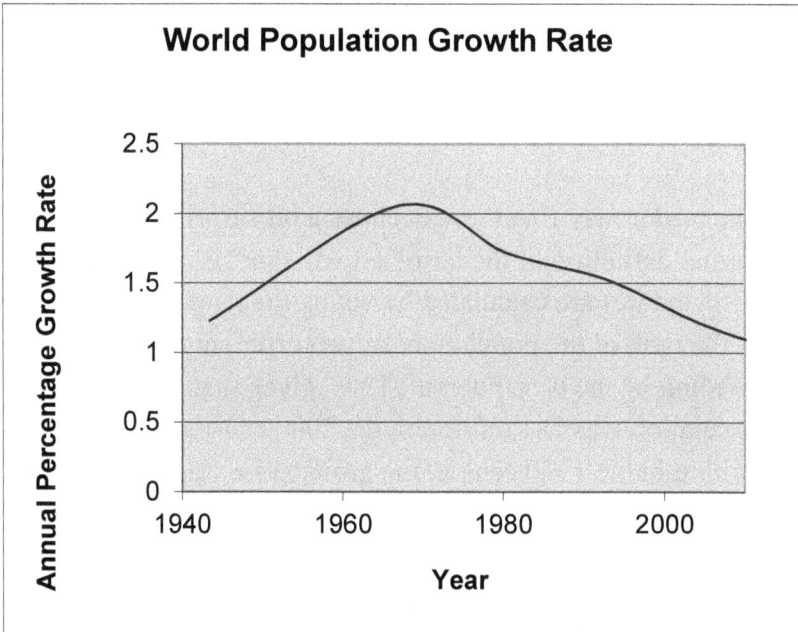

World Population Growth Rate

Figure 11.1 World Population Growth Rate

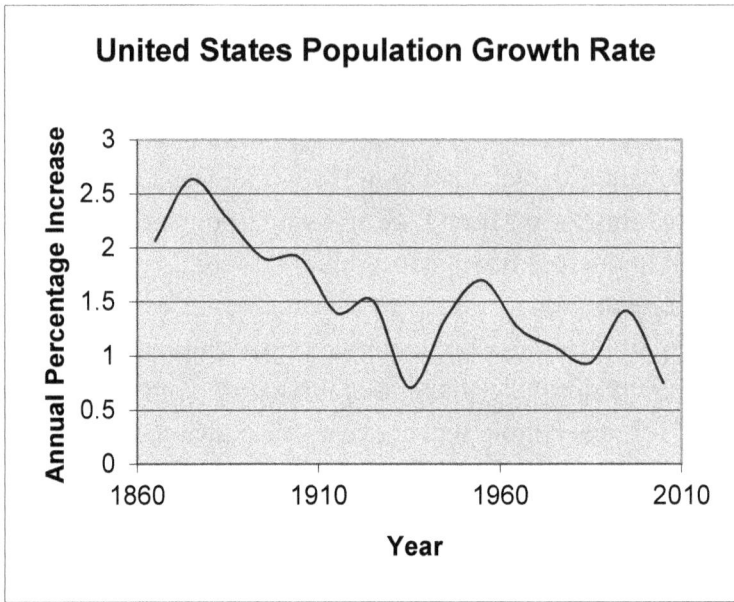

United States Population Growth Rate

(Chart: Annual Percentage Increase vs. Year, with y-axis from 0 to 3 and x-axis from 1860 to 2010)

Figure 11.2 United States Population Growth Rate

The reader may find these declines a bit surprising, so a formal definition of the term "growth rate" is in order. Growth rates are calculated by taking the natural logarithm of the ratio of the populations at two different times and dividing by the time interval. Thus, given that U.S. population was 281 million in the 2000 census and 249 million in the 1990 census, the growth rate was $(1/10) \times \log_e (281/249) = 0.0121$, or 1.21% per year. Using the same formula, a doubling of population in fifty years implies an average annual increase of 1.39%. Of course, the United

States added more people to the population between 1990 and 2000 than in any other decade, but the rate of growth was relatively low by historical standards.

Another way of looking at growth is to look at population on a logarithmic scale. Figures 11.3 and 11.4 show the general patterns. In these figures, I have used a base-10 log scale

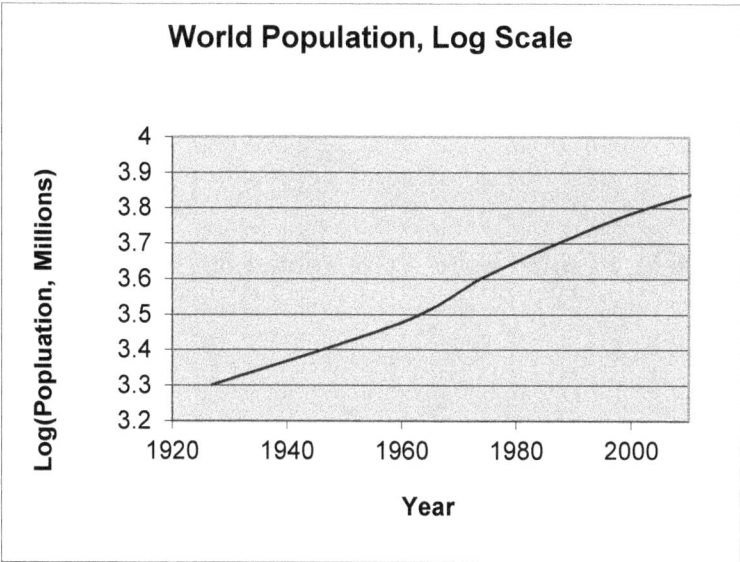

Figure 11.3 World Population Growth

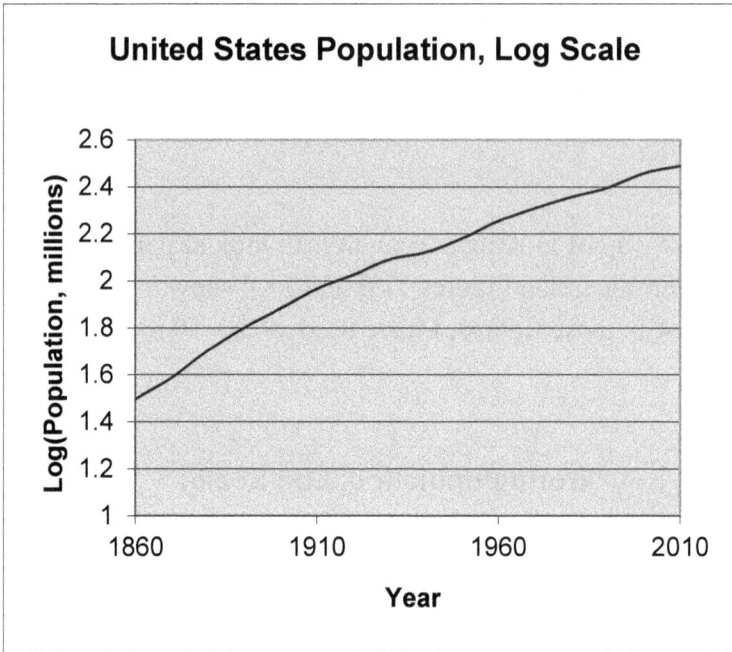

United States Population, Log Scale

Figure 11.4 United States Population, Log (base 10) Scale

because it is easier to remember that the base-10 log of 100 is 2, so that the U.S. population in 1915 was about 100 million, while the world population is approaching ten billion (10,000 million), where base-10 log of 10,000 is 4.

Most residents of our planet are not happy about these growth rates. The world is more crowded, and we can all feel it. The good news, insofar as there is good news, is that the days of doubling in fifty years (or less) are gone. Indeed, there will not be another doubling. World population is expected to increase[19] from 6 billion in 2000 to 8.9 billion in 2050, at which time the growth rate should have declined to 0.4% annually from 1.2% today. In the industrialized world, populations have already begun declining, especially in Europe. By these types of estimates[20], world population should peak at 9.3 billion around the year 2075, and then begin to decline, slowly.

Even countries that have not experienced a rapid increase in population over the past fifty years feel more crowded today. What has changed most is wealth. There are more cars per person, more retired people, more vacation hours. We travel faster, over longer distances. Every beautiful spot on earth has more visitors, far more than a doubling suggests. More than population, it is money that makes it seem the world has less space. More money, a higher standard of living, means more resources consumed per person. There is concern that, as the world becomes wealthier, it will feel far more crowded than another 50% increase in population would suggest. There is an even greater concern that the world will exhaust its supply of non-renewable resources – primarily oil and gas - long before its population begins to decline.

---

[19] J.E. Cohen, Science, Vol. 302, p. 1172 (Nov. 14, 2003)
[20] A simple model for future world population is $\ln (P/P_0) = 0.012t - .000165t^2/2$, where t is years after 2000 and P_0 is 6 billion.

The United States has a well-deserved reputation as the most profligate consumer of non-renewable resources of any nation on earth, although, on a per capita basis, Canadians are worse. In both nations, a majority of people live in moderately harsh climates, with cold winters and hot summers. In both nations people live largely in the suburbs, which are doubly inefficient: Small houses consume more fuel for heating and cooling than high-rises, and suburban living is nearly synonymous with driving a car to work. Cities in North America are spread out over thousands of miles so that business travelers and families think little of going vast distances for meetings or vacations. In the United States, fuel taxes are collected almost solely for the purpose of building and maintaining roads, rather than as a way of raising general revenues. It is no wonder Americans consume so much.

There are two main ways of looking at U.S. energy consumption, per person or on the basis of energy used per dollar of "real" gross domestic product, i.e., GDP corrected for inflation. Energy use per person has been fairly stable, no longer increasing at a rapid rate as in the 1950s and 1960s.

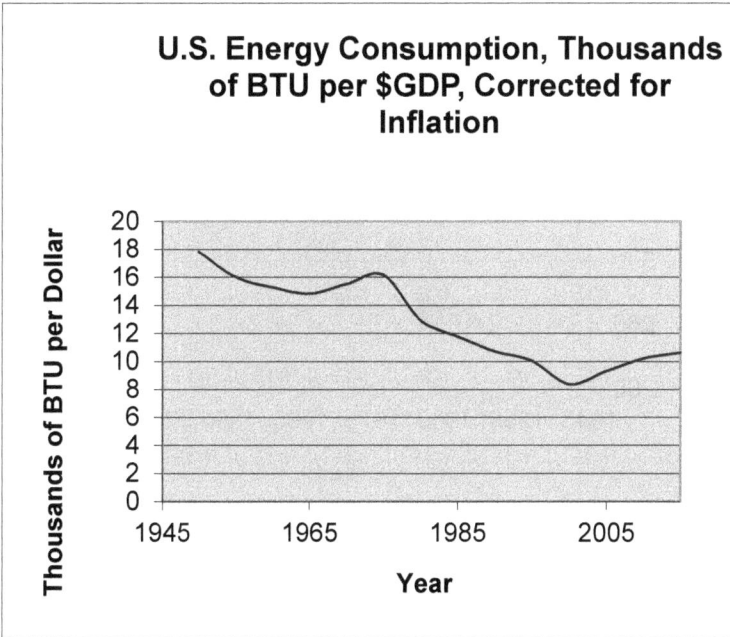

U.S. Energy Consumption, Thousands of BTU per $GDP, Corrected for Inflation

Figure 11.5 United States Energy Consumption per Inflation-Adjusted Dollar of GDP

Energy consumption per dollar of GDP has declined fairly steadily, due primarily to improvements in efficiency. In sum, Americans have become more efficient and more profligate. Their cars get better mileage, but they drive longer distances; their houses are better insulated, but they build bigger houses.

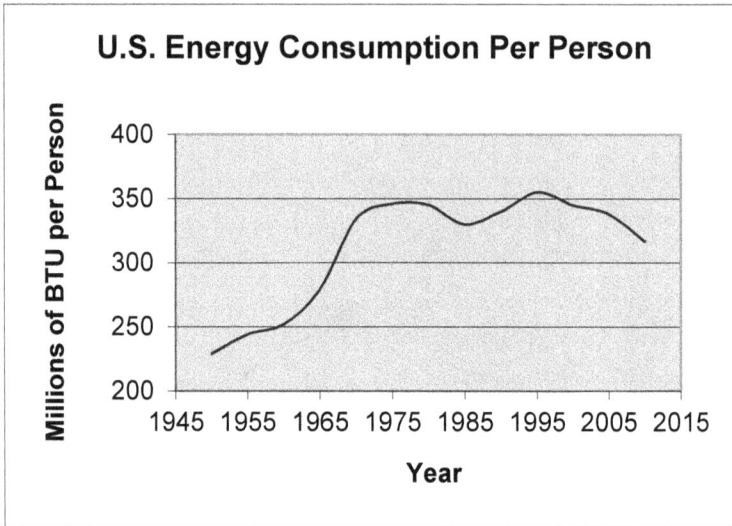

Figure 11.6 United States Per Capita Energy Consumption

To understand how future growth in population and wealth will affect energy consumption it is useful to examine current patterns of U.S. energy consumption. In 2000 the United States consumed 98.5×10^{15} BTU[21]s of energy, of which 39% came from oil, 24% from natural gas, 23% from coal, 8% from nuclear energy and 7% from renewable sources (primarily hydroelectric). This breakdown contains a serious inaccuracy, as it values oil, gas, and coal in the same way that it values electricity. This is, simply put, absurd. It takes roughly 3 BTUs of oil, gas or coal to generate a BTU of electricity. To put these energy sources on a roughly equal

---

[21] The BTU, or British Thermal Unit, a very old-fashioned measure, is the energy required to raise the temperature of a pound of water by one degree Fahrenheit. It is equivalent to 1055 Joules

footing, we should multiply the nuclear and hydroelectric sources by three, in which case the breakdown was 30% oil, 18% each coal and gas, 19% nuclear and 16% renewable. The U.S. exported some coal (1.5 x 10^{15} BTUs) and some oil (2.2 x 10^{15} BTUs), but imported 23.8 x 10^{15} BTUs of oil and 3.8 x 10^{15} BTUs of gas, so net imports were roughly 19% of adjusted consumption, a percentage that is nowadays decreasing. Oil production in the United States apparently peaked four decades ago, around 1970 (11.7 million barrels per day), reached a trough around 2005 (7.9), but has recovered to new heights (12.1) due to fracking. Likewise, gas production no longer seems to be reaching a peak.

How long can these levels of consumption continue? What will happen as the rest of the world becomes increasingly prosperous? Excluding the United States, world energy consumption in 2000 was 52 million BTUs per person per year, about a seventh of the U.S. rate. By comparison, the rate of consumption in Europe was about half the U.S. rate, reflecting the higher population density and somewhat more temperate climate. Most of the population growth over the next fifty years will occur in even more temperate parts of the globe, so no one expects world consumption levels to approach North American and European rates. Yet it is not population that will drive consumption, but wealth, particularly in large nations such as India and China that are still in the early stages of industrialization. Thus, demand for energy can be expected to double or treble before 2050, and to remain at high levels for at least another one or two hundred years. Trying to make economic predictions over such a long period of time is obviously futile – just look back at the technological changes of the past century or two – so it is sensible to focus on the immediate future, the next twenty or fifty years, in which world demand can be expected to grow at a rate of two percent annually, a little higher than the current 1.5%. Some of this growth seems sustainable. Supplies of coal and nuclear energy appear to be adequate to keep up with rising demand. It is oil and gas that are worrisome, particularly if coal burning is reduced for health and environmental reasons.

Consider the situation with respect to oil. In 2000, world reserves were roughly 10^{12} barrels or 5.8×10^{18} BTUs. The world consumption rate was 1.54×10^{17} BTUs annually (in 2000), so reserves were adequate for 38 years at the then current rate, or about 28 years if consumption increased at a 2% annual rate[22]. This sounds a little scary, but it turns out that the phrase "world reserves" does not represent an absolute limit to petroleum sources. As shown below, new reserves are discovered every year[23, 24].

---

[22] World oil consumption is often reported as millions of barrels per day. Since a barrel of oil contains an energy equivalent of 5.8×10^6 BTUs, 75 million barrels per day is 159×10^{15} BTUs per year.

[23] U.S. reserves in 1970 were 39 billion barrels, versus 22 billion in 2000. In the interim, roughly 85 billion barrels were pumped from the ground!

[24] However, discoveries of new oil fields plummeted in the 1990s, so perhaps "world reserves" is not as elastic a phrase as the previous footnote implies. Yet discoveries soared after 2000, due to fracking, although Figure 11.7 indicates they have flattened once again.

World Oil Reserves

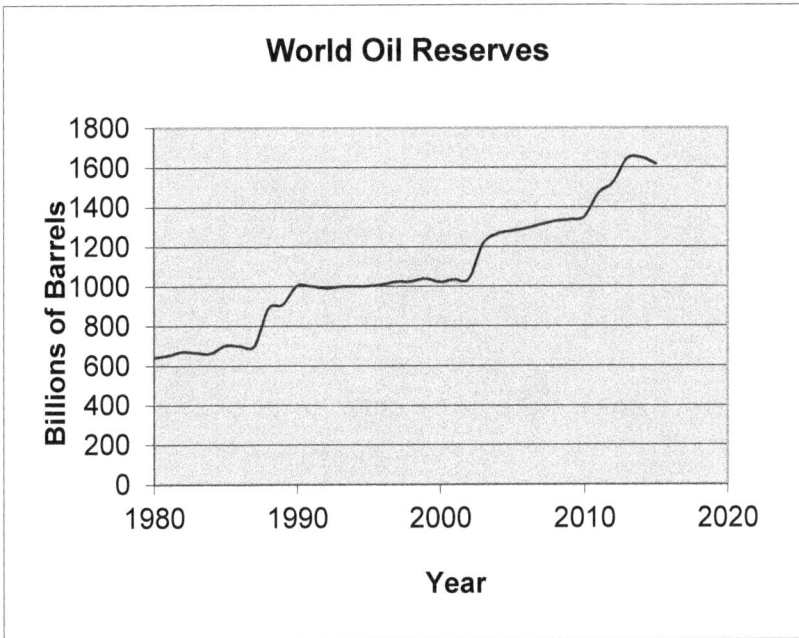

Figure 11.7 World Oil Reserves. Gas Reserves are similar

(https://www.indexmundi.com/energy/?product=oil&graph=reserves)

So it is not completely implausible, on the basis of these numbers, to suppose that oil supplies might remain adequate through the year 2050. However, looking at oil by itself is misleading. World oil production may be approaching a peak. What has kept demand and prices low is the increasing substitution of natural gas for oil. Of worldwide total oil and gas consumption, gas now accounts for 37%. Natural gas production in the United States is about the same as it was thirty years ago. The world is **not** reaching a peak in natural gas production. Does this mean that world oil reserves will last longer? The rate at which oil reserves are depleted is linked to the supply and demand for gas. And gas reserves are even more uncertain than oil. The increasing substitution of gas for oil means additional uncertainty in the rate at which these fuels are depleted. Still, the U.S. Department of Energy does make estimates of worldwide reserves, and in 2000 a good guess was 5.5×10^{15} cubic feet, or, since the energy equivalent of a cubic foot of gas is 1030 BTUs, 5.7×10^{18} BTUs, equal, almost exactly and with misleading precision, to oil reserves. So our estimate of 28 or 38 years of supply of oil and gas is perhaps a bit conservative.

There is another problem with respect to the consumption and production of natural gas: Transportation. Oil can be shipped tens of thousands of miles at low cost. Gas can also be moved long distances in pipelines, but transport over water is relatively expensive: The gas must first be refrigerated (to -160°C) and liquefied. In 2000 the United States had only four ports that accepted LNG (liquefied natural gas), and imports of LNG met only a few percent of U.S. demand. Today there are over 100 LNG facilities. Yet it is widely believed that the explosion of an LNG tanker in port would result in a catastrophic loss of life, so there is much reluctance to expand LNG imports and exports (although there has not yet been an accident involving an LNG tanker, in port or at sea, so the hazards are hard to estimate).

There is yet another question relating to world reserves of oil and gas: Price. In most of the world's oil and gas fields, production is extremely profitable: The costs of finding fields and drilling for oil and gas are a small fraction of the price received. Prices are largely controlled by OPEC, which (somewhat loosely) sets production flow rates for about a third of the world's oil supply. Oil and gas would be much cheaper if competition were more intense. Conversely, production would increase if prices were higher. When we speak of "reserves," we are talking about potential production that is economically feasible at or near current prices. For example, Canada has vast reserves of tar sands that in the past were untapped because prices for oil were not high enough. These reserves are of the same magnitude as the "world reserve" of 10^{12} barrels quoted above.

I hope the reader will understand that the question raised above – how long will energy supplies last – cannot be answered here. Over the past twenty or thirty years, I have read dozens of attempts to answer this question, and time has shown that each attempt has been a failure. On the basis of this short history, I can only say that energy reserves have been underestimated repeatedly. For me, the boy has cried wolf once too often. That doesn't mean the wolf will not soon be at the door. It does mean that estimates of supply and demand are completely unreliable when extrapolated out twenty or more years. (Population estimates are probably much better.)

Twenty years is a reasonable time frame to examine although uncertainties are still huge. I think we can say that oil and gas supplies will remain adequate through that period, at prices that are not far above current levels. Twenty years is a comfortable time frame, as it leaves time for new technologies to be invented and put into use. The cheapest technologies are on the demand side of the economic equation: Improved efficiency of energy use. Cars, trucks, lights, and refrigerators can be designed to consume far less energy while delivering the same benefits. Unfortunately, demand-side improvements cannot usually be made quickly; it takes time for efficiencies to be adopted. But ten or twenty years are generally adequate. The real question is whether supplies diminish and prices rise slowly or rapidly and whether these changes can be foreseen and predicted in advance. Societies and technologies respond well to predictable change; it is the unpredictable that is most disruptive. For the time being, all we can say is that energy prices will start to turn steeply upward some twenty or more years from now, but that is too far in the future to call for much immediate preparation.

Of course, nothing I have said should excuse blind stupidity and profligacy. Many of our "appliances" (cars, refrigerators, air conditioners) consume excessive amounts of energy over their expected lifetimes, excessive in the sense that it makes clear and obvious sense to choose more efficient (and expensive) items that cost less in the "long" run, where "long" means a predictable ten or fifteen years. Governments could help ensure that consumers make choices that are beneficial to all by demanding minimum efficiency standards. Governments could also ensure that, through taxation, prices for oil and gas rise at smooth, predictable rates. This would also help solve another problem peculiar to the rich nations of the world, an aging, and diminishing population. The taxes collected on fuel could help support the needs of the elderly. Unfortunately, governmental foresight is a rare commodity. Killing two birds with one stone is never likely.

Problems and Exercises

Exercise #1: Estimates of future world population vary over a wide range. Find the highest and lowest credible estimates for 2050 and 2075. What reasons do the researchers give for such a wide range? What parts of the world are expected to grow most rapidly?

Exercise #2: Estimates of oil and gas reserves are also quite uncertain. Examine some of the controversies related to making these estimates.

Exercise #3: What fraction of U.S. oil consumption is used for transportation? What fraction is used for residential and commercial space and hot water heating?

Exercise #4: What fraction of U.S. natural gas consumption is used for residential and commercial space and hot water heating?

Exercise #5: Compare the efficiency of a traditional incandescent light bulb with that of an LED bulb. How many hours of use are required to pay for the higher cost of the LED?

Chapter 12: Making Cold: Heat Pumps and Refrigerators

Here on planet Earth, it's easy to make heat. The landmasses are rich in plants, and the atmosphere is rich in oxygen, which, taken together, store the sun's energy in the form of leaves, wood, and an oxidizing agent. Just light a match, and the wood reacts with O_2, producing heat and carbon dioxide (CO_2). We live in an unstable world where only a little push is needed to start a raging fire. If it were not for water to extinguish the fires, this instability would be unsustainable. As it is, plants take CO_2 and H_2O from the air to make the complex carbon chemistry of life, producing oxygen in the process, and later rot or burn to return the CO_2 and H_2O to the atmosphere and, in so doing, take back the oxygen.

To make heat, the initial match is crucial. The plant world, while unstable, does not ignite easily. The balance is delicate, deliciously so: Animals evolved to take advantage of it, to consume plants and digest them slowly, extracting the sun's stored energy slowly rather than in a conflagration. Our bodies burn plant fuel at their own pace.

If making heat is easy, making cold is not. Heat energy flows from hot objects to cold ones; to make heat flow backward requires work. The earth cools off at night by radiating energy into the chill of outer space, but this process won't help us escape the heat of the noonday sun.

The usual way to produce cold is to use a liquid that evaporates rapidly. Generally speaking, the lower the boiling point of the liquid the faster it will evaporate. Liquids evaporate even at temperatures well below their boiling points, provided that the vapor is able to escape. This is why water produces a cooling effect at ordinary room temperatures even though its boiling temperature is 100° C. But this works only if the air is dry to begin with; water won't evaporate into moist, humid air. But in dry climates evaporating water will produce a chill, just as a thin coat of water on your finger will cool it down if there is just a little dry breeze to blow away the vapor.

Modern refrigerators and air conditioners use a pressurized liquid, typically one of the so-called freons, to produce the chilling effect. Freon vapor is fed to an electrically driven pump, where it is compressed to a much higher pressure, so high that the vapor starts to condense and form a liquid. A condensing gas produces heat, and this heat has to be disposed of. Typically, the heat is rejected to someplace we don't mind heating up, your kitchen, for example, if we are talking about a refrigerator, outdoors, if we are talking about an air conditioner. The liquefied gas, under pressure, is cooled down a bit. The resulting liquid is then bled to an evaporating unit, where it is allowed to boil at a much lower pressure. The evaporating unit becomes chilled, and the low-pressure gas is fed back to the compressor pump, thus completing the cycle. In this fashion, the electrically driven compressor pump produces cold in one place (the evaporator) and heat in another (the condenser). Heat is pumped "uphill," from a cold place to a warmer place.

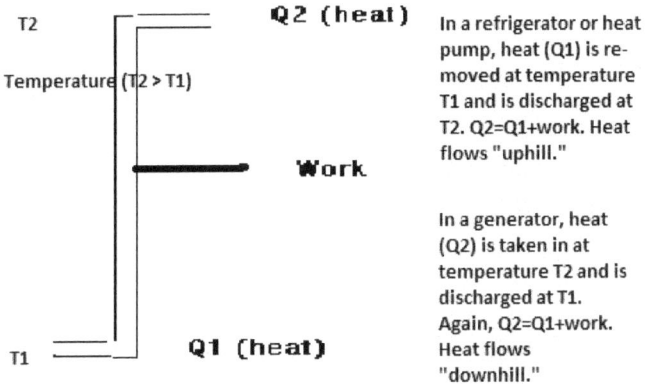

T2

Temperature (T2 > T1)

Q2 (heat)

Work

T1

Q1 (heat)

In a refrigerator or heat pump, heat (Q1) is removed at temperature T1 and is discharged at T2. Q2=Q1+work. Heat flows "uphill."

In a generator, heat (Q2) is taken in at temperature T2 and is discharged at T1. Again, Q2=Q1+work. Heat flows "downhill."

Figure 12.1 Schematic energy flows for refrigerators, heat pumps, and generators

The laws of thermodynamics put strict limits on the efficiency of a heat pump. If T_2 is the (absolute) temperature of the condenser, T_1 is the temperature of the evaporator, w is the work (energy consumed per unit of time, e.g., watts), and Q_1 is the rate at which heat is removed from the evaporator, then the efficiency $\varepsilon = Q_1/w$ can never be greater than

$$\varepsilon < T_1/(T_2 - T_1)$$

At first glance, this is not much of a limit. For example, in an air conditioner, if T_1 is 10° C, or 273 + 10 K in absolute terms, and T_2 is 40° C, or 273 + 40 K in absolute terms, then the limit is $\varepsilon < 283/30$, or 9.4, 940%! Some limit! For each Joule used by the compressor pump, 9.4 Joules of heat can be pumped uphill. Furthermore, the temperature limits we have chosen are reasonable: 10° C is 48° F, quite chilly air, while 40° C is 104° F, quite a hot day outside.

Real air conditioners don't do anywhere near this well. Air conditioners are rated in an odd way, by their EER, which is Q_1 expressed in BTU per hour and w expressed in watts. For modern room air conditioners, a rating of 11 is considered good. To convert EER to efficiency we note that a BTU is 1055 Joules, so 11 BTU/hr is 11 x 1055/(60*60) = 3.2 Joules/second, or 3.2 watts, an efficiency of 320%. So air conditioners are only about 35% as efficient (3.2/9.4) as the theoretical limit. Why?

First, fans are needed to move the air past both the evaporating and condensing units. The energy used by the indoor fan adds to the heat to be removed, Q_1, and the energy used by the outdoor fan is simply lost. Second, there are temperature drops in the heat exchangers in both the evaporator and the condenser: 10° C may be chilly indoor air and 40° C may be hot outdoor air, but another 5°-10° C on both sides are generally required so that heat is removed more easily from the air passing by the evaporator and so that heat is rejected more easily to the air passing by the condenser. This is basically a tradeoff between the size and energy consumption of the fans and the losses incurred by going to a larger temperature difference between the condenser and evaporator. The effect is to change the theoretical efficiency limit from 283/30 to 278/40 or 273/50, from 9.4 to 6.9 or 5.5. Finally, the compressor motor is not 100% efficient in converting electrical energy into useful work, and some energy is lost here as well. So an overall efficiency of 320%, 35% of the theoretical limit, is not entirely surprising.

Can we do better? Of course. Compressors and fans can be made more efficient; heat exchangers can be built with more surface area, to reduce the temperature drops between air on one side and liquid on the other. Are the added costs worthwhile? It depends on where you live. If your air conditioner is used only on the hottest days of the year for a few hours each day, paying for improved efficiency is pointless. If your air conditioner runs around the clock for months on end, it probably makes sense to spend more money up front. When it comes to air conditioning, one size, one efficiency, does not fit all needs.

So far I have focused on the air conditioning unit itself, ignoring inefficiencies that are part of the space to be cooled. Better home insulation will reduce the cooling load. Window placement is crucial: North- and south-facing windows do not let in much solar radiation in summer; east- and west-facing windows are a disaster. Drapes help, but well-placed trees and shrubbery on the east and west sides of a house are even better. By shielding a house from direct solar radiation, vegetation can reduce the cooling load dramatically, although not all of the load is simply heat transmitted through walls and windows. In humid climates, a good portion of the cooling load is attributable to dehumidification, removal of water vapor from the air. The efficiency of this process depends mostly on how well a house is sealed against outside air and humidity.

There are other inefficiencies in the space to be cooled. Inefficient electric lighting and inefficient refrigerators generate more heat than needed, and this heat adds to the cooling load. Replacing incandescent lamps with fluorescents and LEDs not only consumes less electricity for lighting but also reduces electricity consumption for cooling. Managers of large buildings are generally aware of these tradeoffs; homeowners are less likely to understand the benefits of these simple conservation measures.

Similar tradeoffs are made in the design and manufacture of home refrigerators and freezers: Compressor efficiency, heat exchanger size, and insulation. In a superbly insulated refrigerator (think of a thermos bottle), the only large losses should occur when the door is opened. Unfortunately, this statement is not quite true, particularly in freezers, where the temperature is typically $-15°$ C, well below the freezing point of water. Any water vapor inside the freezer will inevitably form ice on the surfaces of the heat exchanger, and ice is, as any Eskimo can attest, a good insulator, thus reducing the heat exchanger efficiency in the evaporator unit and increasing the temperature difference needed to remove heat effectively. In modern "frost-free" freezers, ice buildup is countered by heating the evaporator heat exchanger surfaces periodically, melting the ice and removing the water. This is a "waste" of energy, but less of a waste than allowing the ice layer to build up. As is the case with air conditioners, no single refrigerator design is optimal for all climates and uses, although the range of sensibly efficient possibilities is smaller. For example, the 12V fridge I use in my (solar powered) summer house has separate compressors for the freezer and fridge. This is a very unusual design, but it does provide much higher efficiency than a conventional one-compressor system and allows me to use a smaller solar panel to provide power.

Air conditioners can also be used for heating if they are run in reverse as heat pumps. The theoretical efficiency limit for this application is

$$\varepsilon \;<\; T_2/(T_2 - T_1)$$

which is, as before, greater than 100%. This is not a figment of scientific and technological imagination. Electricity can be converted into heat at better than 100% efficiency, simply by removing heat from the cold outdoors and pumping it uphill to a higher temperature. The major obstacle to widespread use of heat pumps is frost build-up. In a freezer, frost is limited by evaporation of water from the stuff that has been frozen, and from the humidity in the air that leaks into the freezer when the door is opened. In a heat pump, outdoor air is continuously blown across the surfaces of the (outdoor) evaporator heat exchanger, so that ice build-up is a constant problem. Heat pumps are particularly useful if the local climate is arid so that the humidity is low, or mild, so that ice formation is rare, but in more humid and colder climates the frost problem can be solved by briefly running the heat pump in reverse to melt the frost. In some locations, the evaporators can be placed underground where it is warmer, or where there is a source of (relatively warm) water, which is a far better medium for heat exchange than outdoor air. In the right climates, heat pumps are a very attractive and practical way to heat a building, and the same system can be used for summer cooling, thus reducing overall capital costs.

From the point of view of thermodynamic theory, a heat pump or a refrigerator or an air conditioner is simply an electric generator running backward. When fuel is burned to make heat, energy flows from the hot heat source to the cold environment, and a portion of this heat can be captured to make electricity. Thermodynamic theory tells us that the maximum efficiency for this process, $\varepsilon = w/Q_2$, is

$$\varepsilon = (T_2 - T_1)/T_2 = 1 - T_1/T_2$$

which is inevitably less than 100%. Theoretical efficiency depends on two factors, the "flame temperature" T_2, the temperature at which heat is generated, and T_1, the temperature at which heat is rejected to the environment. Fuels such as natural gas have very high flame temperatures, typically over 4000 K, so the theoretical limit can easily exceed 90%. However, steel melts around 1500° C, so the flame temperature is, in the end, irrelevant. The upper temperature limit is actually set by the properties of water, which is used as the evaporating and condensing liquid. Above 373° C, known as the critical point, water will not condense to a liquid, no matter how high the pressure. It is far more expensive to build a heat exchanger to extract heat from a gas than from a liquid. So 373° C, or 646 K, is the practical upper temperature limit T_2.

The lower limit T_1 depends on how the heat is rejected, whether to a lake or river or to the atmosphere, in a cooling tower. For example, assume T_1 is 300 K (27° C, 81° F). The theoretical efficiency limit tumbles 53.6%. In practice, power-plant efficiencies are around 35% to perhaps 40%, a long way from the theoretical 90+%. As with air conditioners and heat pumps, the practical efficiency of an electric generator is only about 35% of the theoretical limit.

Generators based on natural gas (methane) are not limited by the properties of water but are based on a different principle altogether. Like automotive engines, they rely on the expansion of combustion gases when a fuel is burned. Nevertheless, natural gas turbine generators are rarely more than 35% efficient, although, with advanced technologies, there is good reason to believe their efficiencies could be pushed to 50% or more. The main attraction of gas turbines is that they are relatively cheap (and quick) to build, are less polluting than coal or oil, use a fuel that has (until recently) been inexpensive, and can be turned on or off quickly, so that electrical supply and demand can be matched.

Finding ways to improve upon these efficiencies is an important engineering challenge, but it is hardly a new challenge. The problem has resisted attempts to find mold-breaking solutions for well over a hundred years. Much effort today is focused on fuel cells, which convert chemical energy to electrical energy by slower "electrochemical" processes that are similar in principle to the processes that animals use in consuming food. Hydrogen (H_2) and methanol (CH_3OH) are presently the fuels of choice. Progress is slow, but the potential promise is huge, so the research effort is active and well-funded. Still, I bear in mind something that a colleague active in fuel cell research told me twenty-five years ago. "Fuel cell researchers say that fuel cells will be ready for widespread commercial use in twenty years. The problem is they said the same thing twenty years ago." It still hasn't happened, although this time I am tempted to believe the researchers.

There are other ways of converting heat (or chemical energy) to electricity. For example, solid-state devices, known as thermoelectric generators, are widely used in deep-space probes, with a radioactive core supplying the heat. They work quite reliably, but their efficiencies are atrocious. Other approaches, involving biotechnology, imitating the processes of living organisms, are surely worth examining. I am not optimistic, not in the short run. As for the long run, anything is possible within the limits that thermodynamics imposes. As John Maynard Keynes famously reminded us, in the long run, we are all dead. However, efficiency problems will survive us, and they will surely be important to our children and grandchildren as supplies of fossil fuels inevitably diminish or are restricted for health or environmental reasons.

Problems and Exercises

Exercise #1: My 12V refrigerator-freezer uses only 0.8 kilowatt hours per day. What is the comparable electricity consumption for a typical 120 VAC unit? (Assume a useful capacity of 16 cubic feet freezer and refrigerator space.)

Exercise #2: The pump that supplies water to my summer house can pump 100 gallons per hour a height of 100 feet when powered by 2.8 amperes at 24 volts. (a) Convert the U.S. units to their metric equivalents. (b) Calculate the efficiency of the pump.

Exercise #3: From published sources, find out what is the typical efficiency of a refrigerator compressor.

Exercise #4: In a power plant that generates electricity, the heat from burning coal or oil or gas that isn't converted to electricity is "waste heat," that is, it is discharged to the environment (air or water). Could this heat be used instead for residential and commercial space heating in winter? Has this been done? What are the obstacles?

Exercise #5: In New England, a house uses about 1000 therms of natural gas per heating season. (A therm is

100,000 BTU, and is the heat released by burning about 95 cubic feet of gas.) However, about 25% of the heat goes up the chimney, due to a variety of losses. Alternatively, one can stay warm with electric baseboard heaters, which are 100% efficient, but electricity is more expensive than gas, roughly $0.10 per kilowatt-hour compared to about $1.70 per therm for gas. Thus, the cost of gas is about $1700 per season, while the cost of electricity would be about 0.75 x 0.10 x 10^8 x 1055/(1000 x 60 x 60), where 1055 is the number of joules in one BTU. This works out to about $2200. If instead of baseboard heating an electric heat pump were used, how high would its efficiency have to be to make electricity competitive with natural gas?

Chapter 13: Computers and Mathematical Models

Science advances in three ways: New theories, new observations, and new tools. Not that these three can be separated so cleanly. New tools lead to new observations; a new theory **is** a new tool. But a cheer or two or three is in order for the toolmakers, whose contributions to science are often overlooked. Science makes progress because scientific progress leads to new tools; science begets technology which begets new science.

The single most important tool of the last fifty years is the digital computer, specifically the microprocessor, which is now ubiquitous in our cars, telephones, CD players and televisions. Vacuum tubes have nearly vanished.

The first computer I ever used was the Edsac II, and, as the suffix –ac implies, it was based on vacuum tubes, not transistors. I don't think it ever ran for more than an hour before something broke, requiring replacement of what today we would call a printed circuit board, except that the things must have weighed hundreds of pounds apiece and were set on rollers to make replacement easier. There was something romantic about the old clunker, with a loudspeaker attached to the innards so you could hear what it was up to, and if necessary abort a program if it sounded like it had entered an endless loop. A fictionalized version[25] appears in Fred Hoyle's science fiction novel "The Black Cloud," although Hoyle's machine was considerably more reliable. Edsac II died in early 1964, electrocuted, I think, by 220 volts in the wrong circuit, and the Cambridge University Computation Center shut down until Edsac could be replaced by a solid-state machine.

Edsac II had less computational power than one of today's $100 programmable calculators. Its applications were all mathematical: In those days, computers calculated. Edsac did have one novel feature, a plotter that Cambridge astronomers used to map the intensity of radio-frequency stellar emissions; this was in the days when pulsating, beacon-like stars, pulsars (also known briefly as LGMs, for "little green men") were first discovered.

The computer on my desktop today, on your desktop, too, is far more powerful than any multi-million dollar university mainframe of thirty years ago: More memory, more speed, more reliability. We are all in the midst of a revolution that has been absorbed so quietly that few of us are properly astonished and amazed. Once, computers calculated; nowadays, they play music and videos, read texts, understand (or at least convert to text) human speech,

---

[25] Fred Hoyle, "The Black Cloud," Harper & Row, New York, 1957.

translate languages, search for information, and connect people together in ways completely unforeseen just ten years ago.

Scientists like to say that they stand on the shoulders of those who came before them. Nowhere is this truer than in the world of computers and computer software. Each generation of programmers refines and enlarges the software of the past, not only improving speed and ease of use but also by making the software more modular so that sections of code can be linked together in novel ways. In the early days, computer operating systems did little more than process input, display text output on a screen or printer, and store and retrieve information from a disk drive; nowadays, operating systems have many thousands of capabilities, from formatting text and pictures to speaking simple sentences.

When I first began using computers, scientists wrote their own programs. Every computer had, in its operating system, a language designed to facilitate computation, usually FORTRAN or Algol. If you wanted to fit data to a straight line (or other function), you wrote your own least-squares program. If you wanted to multiply matrices together or invert them, you wrote your own code. Some programs were published in specialized journals; others were traded among graduate students. Professional programmers at university computer centers and computer manufacturers developed standardized packages for common applications, such as statistical analysis, and professional organizations developed standards for the common "high-level" computational languages such as FORTRAN and COBOL. Computer users were a mix of self-taught amateurs and amateurs who had become professionals; the field called computer science was not widely taught.

History repeated itself when desktop computers were invented in the early 1980s. The operating systems were simple, but usually included a high-level programming language such as BASIC or Pascal. Once again users wrote their own programs. Once again, standardized packages were developed for common uses. This time around the number of amateur programmers exploded, and popular magazines sprung up to serve the new community with advice, software samples, and advertisements. The biggest applications of the day were games, although, for serious uses, there were simple word processors, spreadsheets, and database managers. Once again, some amateurs became professionals; I was one of them. This time around the number of professionals grew at an extraordinary rate because so many more people were exposed to computers

If you purchased a computer in the last five years – and you probably did, because computers get so much better and cheaper every year – it probably did not come with a real, general-purpose programming language. The revolution is over. If you want to write programs, you'll probably have to buy the software separately, if you can find it, or take a programming course. No more amateurs, except for high-school students. For some purposes, the programming languages built into spreadsheets such as Microsoft Excel are adequate for learning how programs are written and developed, but as programming languages go, they are limited in scope and hard to understand. But they are all the typical computer user is likely to own.

This is a situation I much regret. General-purpose programming languages are one of the few tools of science that are, or ought to be, free and available to everyone. Anyone interested in science, anyone who wants to understand scientific modes of thought, should be able to write simple programs. It is a useful skill, much like using an electric drill or saw; you need not be an expert, or even very good, to find value it in. And rudimentary programming can be learned in a matter of days if you are logically or mathematically inclined. It is also true that to be a really good programmer takes years of study and more years of experience. A good programmer is like a master cabinetmaker: You talk to your customers, you listen to their needs and desires, you steer them in the direction of what is practical and cost-effective, and you execute a mutually agreed-upon design. But you don't have to be a master to use simple tools for your own benefit. I recommend downloading a copy of GWBASIC[26], BASICA or QBASIC (Quick Basic) for your own use; these programs run under MS-DOS in Microsoft Windows.

 Let's examine a few of the problems raised in previous chapters that are amenable to simple software solutions. In Chapter 3 I asked how ancient astronomers calculated the intervals between eclipses and wrote down the following equation that should predict the time interval between repetitions of a given type of eclipse:

$$29.5306 \; n_1 = 27.2122 \; n_2$$

where n_1 and n_2 are the numbers of Synodic and Draconic months respectively. I noted that this equation can be solved by trial and error and a simple computer program will do the job:

```
10    DEFDBL A-Z              REM
      Use double-precision
20    F1=29.5306             REM
      Days in a Synodic month
```

```
30      F2=27.2122              REM
        Days in a Draconic month
40      FOR N1% = 1 to 1000     REM
        % means N1 is an integer
50      N1D=N1%                 REM
        Convert the integer to double
60      FOR N2% = 1 to 1000     REM
        N2 is also an integer
70      N2D=N2%                 REM
        Double-precision
80      LHS=F1*N1D              REM
        LHS is the left-hand side
90      RHS=F2*N2D              REM
        and RHS the right-hand side
100     DIFF=ABS(LHS-RHS)       REM
        DIFF is the difference
110     IF DIFF>0.1 GOTO 130    REM  If
        the difference is big, ignore
120     PRINT N1%,N2%,LHS,DIFF REM
        Print the results
130     NEXT N2%                REM
        End of the loop for N2%
140     NEXT N1%                REM
End of the loop for N1%

150     STOP                    REM
We're done
```

This is a pretty stupid program. It performs millions of calculations (1000 x 1000) while a well-written program could do the same job by doing only a few thousand. Who cares? The program gets the job done (it finds the solutions

$n_1 = 223$, $n_2 = 242$ and $n_1 = 669$, $n_2 = 726$) in only a few seconds. It's short, easy to read and to write, and it works.

In Chapter 11 I noted that worldwide oil reserves in 2000 were approximately 5.8×10^{18} BTUs while annual consumption was 1.54×10^{17} BTUs. Dividing reserves by consumption shows that reserves should last 37.7 years if consumption remains constant. But what if consumption increases at a 2% annual rate? A simple program will provide the answer:

```
10      RESERVES=5.8                          REM
        Worldwide reserves
20      ANNUAL=0.154                          REM
        Annual consumption
30      RATE=0.02                             REM
Rate of consumption increase

40      FOR I%=1 TO 40                        REM
For the next 40 years

50      RESERVES=RESERVES-ANNUAL REM
        Decrease reserves by annual
60      ANNUAL=ANNUAL*(1.0+RATE) REM
Increase annual consumption

70      IF RESERVES>0 GOTO 100      REM
Skip if any reserves remain
```

```
80     PRINT"Reserves almost lasted ";I%;" years."
REM  Print the result

90     STOP                          REM
       We're done
100    NEXT I%                       REM
End of 40-year loop
```

The program stops at I%=29, indicating that reserves will last nearly 29 years if consumption increases at the assumed rate.

Problems like these occur in real life: Mortgage payments, investment alternatives, and so on. One common question is whether to begin drawing on social security at age 62 or age 67, assuming you retired early and can afford not drawing right away. The question boils down to how long you have to live before you come out ahead by delaying. However, you shouldn't just add up the dollars. Money received today is more valuable than money received later, since you could invest money received today and withdraw it later. Of course, you'd have to pay taxes on the investment income, which will reduce the effective rate of return. And you have to account for the effects of inflation on your investment since social security payments are adjusted for inflation. Let's suppose that money invested today can earn a 2% rate of return after taxes and inflation. Here's the program, based on a projected[27] social security income of $1000 a month at age 65 or $800 a month at age 62. To make the two projections equivalent, I have assumed that all money from social security is reinvested, not spent. The analysis follows the total investment on a month-by-month basis:

```
10    SUM62=0                              REM
      Starting sum for age 62
20    SUM65=0                              REM
      Starting sum for age 65
30    INC62=800                            REM
      Monthly income for age 62
40    INC65=1000                           REM
      Monthly income for age 65
50    AFTERTAXRATE=0.02                    REM
      Yearly after-tax investment
60    AFTERTAXRATE=AFTERTAXRATE/12
      REM On a monthly basis
70    FOR I%=1 TO 360                      REM
Examine up to 30 years

80    SUM62=SUM62+INC62                    REM
Save the month's income

90    SUM62=SUM62*(1+AFTERTAXRATE)
      REM Earn interest on savings
100   IF I%<37 GOTO 180                    REM
No income until I=37, age 65
```

---

[27] Social security income is reduced by approximately five-ninths of one percent for each month of early retirement. In 36 months, this adds up to 20%.

```
110    SUM65=SUM65+INC65              REM
Save the month's income

120    SUM65=SUM65*(1+AFTERTAXRATE)
REM Earn interest on savings

130    IF SUM65<SUM62 GOTO 180       REM
Don't report yet

140    RI=I%                         REM
Convert integer to 'real' numb.

150    AGE=62+RI/12                  REM
       Calculate age
160    PRINT"Amounts are equal at age ";AGE
       REM Report result

170    STOP                          REM
       We are done
180    NEXT I%                       REM
Loop up to 360 months=30 yrs
```

The program contains a slight flaw, in that interest is earned on the first month's social security check immediately, rather than at the end of the month. This is a trivial error and could be corrected easily, but at the expense of some clarity and simplicity.

If you run this program, it will tell you that you will come out ahead retiring at 65 only if you live past age 79. If you assume a higher rate, say 4% (good luck!), the break-even point is nearly 83. So your decision should depend on your sex, your health and habits, and the longevity of your parents. There are other factors, of course. If you die before your spouse and he or she has lower projected social security earnings, your spouse can take over your higher rate of income. And of course, everything depends on your financial situation, on whether you can afford to delay.

The programs I have outlined above are crude, rudimentary and useful. They provide solutions to simple questions that can be modeled in mathematical terms. By studying, using and expanding upon these programs, readers can teach themselves how to use an important scientific and mathematical tool. The modern alternative, of course, is to log onto the web and find answers by reading what others have said, to leave the thinking and analysis to someone else. The web is often a good place to start. But it is only a start. There is no substitute for your own judgment and analysis.

Problems and Exercises

Exercise #1: Rewrite the eclipse program so that it takes far fewer steps.

Exercise #2: How long will oil reserves last if consumption increases at a 3% annual rate?

Chapter 14: Molecular Models

About a hundred and fifty years ago, chemists came up with some simple rules that led to an extraordinary flowering in the field of organic chemistry. First, they noted that the element hydrogen (H) forms only one bond, that is, it can bind to only one other element at a time. Oxygen (O), on the other hand, can form two bonds, so it can bind to two other elements simultaneously. Indeed, oxygen always forms two bonds. Similarly, nitrogen (N) can form three bonds, and almost always does, and carbon always forms four. When we say "always does," we include the possibility that oxygen and nitrogen and carbon can form so-called "double bonds," that is, two bonds at once to just one other element, and that nitrogen and carbon can form "triple bonds." (Carbon could by this logic make a quadruple bond to another carbon atom, but it doesn't!) Thus, hydrogen atoms can bond to each other and form "molecular hydrogen," H_2 (or H-H), and likewise oxygen atoms can bond to each other (via a double bond) to form "molecular oxygen," O_2 (or O=O), and nitrogen atoms can bond to each other (via a triple bond) to form "molecular nitrogen," N_2 (or N≡N). Indeed hydrogen and oxygen and nitrogen are found in nature as molecules with two atoms.

These rules have many implications. For example, the simplest combination of hydrogen and oxygen is H_2O, water, perhaps more properly written as H-O-H. Hydrogen peroxide also obeys the rules: H-O-O-H. So does carbon dioxide, CO_2, or O=C=O. And methane, CH_4, and ammonia, NH_3. But carbon monoxide, C=O, breaks the rules unless of course, it is really C_2O_2 or O=C=C=O. But it isn't. The attitude of chemists to discrepancies of this sort was to ignore them, to push ahead regardless. The simple rules are not always correct.

How did chemists know that carbon monoxide isn't really dicarbon dioxide, O=C=C=O? The breakthrough was Avogadro's hypothesis in 1811, which stated that equal volumes of gas (at the same temperature and pressure) contain equal numbers of molecules. A liter of hydrogen gas at room temperature and pressure weighs about 0.08 grams, while a liter of oxygen weighs about 1.3 grams and a liter of nitrogen weighs about 1.2 grams. (This means the relative weights of hydrogen to oxygen to nitrogen are 1:16:14, or, taking the hydrogen atom as 1, 2:32:28.) A liter of carbon dioxide weighs 1.8 grams, and a liter of carbon monoxide weighs only 1.2 grams, about the same as N_2, nitrogen. If carbon monoxide were C_2O_2, a liter would weigh 2.3 grams. Case closed, although that means that carbon monoxide breaks the rules unless it is an exception to Avogadro's hypothesis. It isn't.

These rules led quickly to the idea of chemical structure, the notion that chemists could in some sense draw what molecules looked like, at least in stick form. Given, for example, the formula CH_4O, the only structure that fits the rules is H_3COH, or

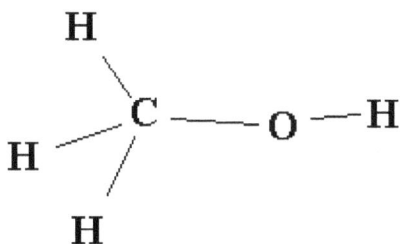

Figure 14.1 Stick Diagram of the Methanol Molecule

although nowadays we tend to write CH_3OH, it being understood that three of the hydrogen atoms bond to the carbon atom while the fourth hydrogen atom is bound to oxygen, which in turn forms its other bond to carbon. What about C_2H_6O? Whoops! There are two structures that fit the rules, namely CH_3CH_2OH, and CH_3OCH_3. And indeed there are two known substances of composition C_2H_6O, ethanol, or grain alcohol, the kind you can drink, and methyl ether. Chemists quickly decided that ethanol had the CH_3CH_2OH structure while methyl ether was CH_3OCH_3. Why? Because ethanol has properties a lot like wood alcohol, or methanol, which they knew was CH_3OH because, as we noted above, this is the only structure that fits the rules. Methanol and ethanol both mix with water in any proportion; methyl ether does not, but forms a separate layer. That's nearly all the evidence that chemists needed to choose one structure over another.

Organic chemists quickly built up a catalog of "functional groups," groups of atoms in a structure that give rise to similar physical properties. For example, anything with an oxygen atom between two carbon atoms, as in methyl ether, has in some sense properties much like methyl ether itself, so we call the compound $CH_3CH_2OCH_3$ "methyl-ethyl ether" (or, in more modern language, propyl ether) and we call $CH_3CH_2OCH_2CH_3$ "diethyl ether" (or just plain ether). Likewise, we call almost any compound that has a C-O-H linkage "an alcohol," regardless of how many other things are attached to the carbon atom. There is one notable exception: If the carbon atom is bound both to the O-H and to another oxygen atom via a double bond, we call it a carboxylic acid. For example, acetic acid has the structure CH_3COOH, or

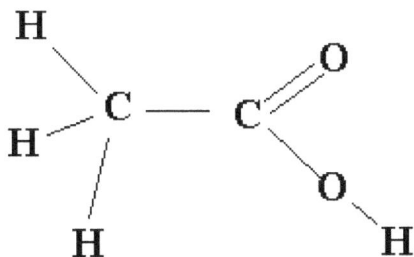

Figure 14.2 Acetic Acid

This compound, and similar compounds containing the - COOH functional group, do not have properties similar to alcohols, despite the -OH linkage, but instead act like acids. This means that they dissolve readily in water and form ions in which the hydrogen atom attached to the oxygen breaks off, forming the ion H^+, leaving behind a negatively charged carboxyl ion.

Compounds that contain the C=O functional group without an adjacent OH fall into two categories, the ketones, such as acetone, CH_3COCH_3, and methyl ethyl ketone (or butanone), $CH_3CH_2COCH_3$,

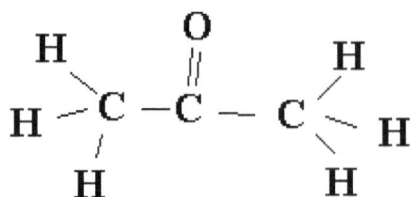

Figure 14.3 Acetone

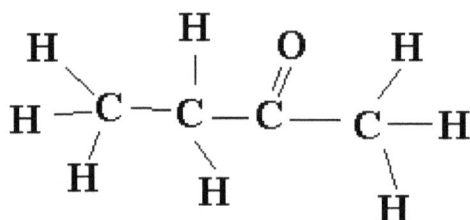

Figure 14.4 Methyl Ethyl Ketone (MEK), aka Butanone

and the aldehydes, such as formaldehyde, HCHO, acetaldehyde, CH_3CHO, and propyl aldehyde, CH_3CH_2CHO:

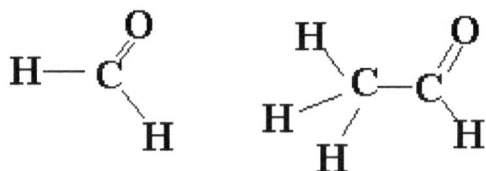

Figure 14.5 Formaldehyde and Acetaldehyde

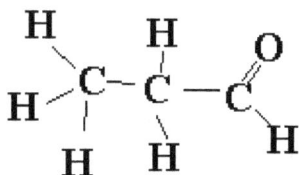

Figure 14.6 Propionaldehyde

In the ketones the carbon that forms a double-bond to the oxygen is also attached to two other carbon atoms, while in the aldehydes the carbon is attached to at most one other carbon atom; the final bond is to a hydrogen atom.

So far we have mentioned five types of functional groups consisting solely of carbon, oxygen, and hydrogen: alcohols, ethers, carboxylic acids, ketones, and aldehydes. This catalog just scratches the surface, for we have not considered compounds containing nitrogen, nor compounds containing multiple carbon-carbon double bonds. It is enough to say that by combining the idea of chemical composition with the notion of functional groups, chemists were able to draw at least hypothetical structures for some very very large molecules long before anyone had any real idea what atoms and molecules were made of. As it turns out, these structures were almost never wrong: Modern evidence rarely contradicts the old structures. This is extraordinary, for as the number of atoms grows, the number of possible structures increases wildly. Consider, for example, acetone, C_3H_6O, which we have drawn as CH_3COCH_3. There are three other

211

structures of composition C_3H_6O that fit the rules: Allyl alcohol, or $CH_2{=}CHCH_2OH$, propylene oxide, which has a ring structure, and propanal, CH_3CH_2CHO, an aldehyde. We have omitted $CH_2{=}COHCH_3$, the "enol" form of acetone, which evidently does not exist, or, more correctly, is believed to be unstable with respect to acetone, to which it can convert by the movement of a single hydrogen atom. This is called the keto-enol isomerization.

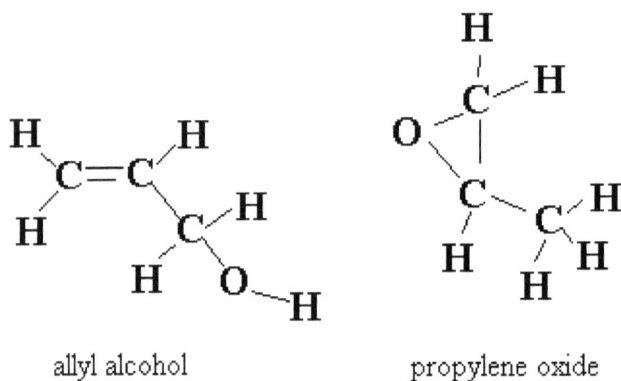

allyl alcohol   propylene oxide

Figure 14.6 Allyl alcohol and propylene oxide

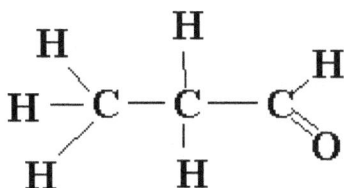

propanal

Figure 14.7 The Structure of Propanal

How do chemists know that acetone is CH_3COCH_3 and not $CH_2=COHCH_3$? It is, in fact, difficult to know which form is correct since any ketone can be written in either form.

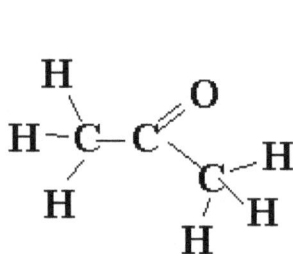

acetone, ketone form

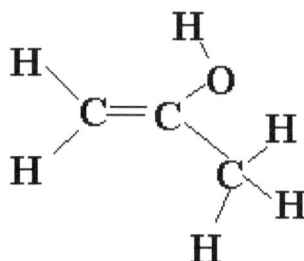

acetone, enol form

Figure 14.8 Keto and Enol Forms of Acetone

213

Acetone has a much lower boiling point (56°C) than allyl alcohol (96°C), which suggests it is not an alcohol, and its density is the lowest of the C_3H_6O compounds, which suggests it does not have a C=C double bond.

Chemists went further than stick structures. They developed the notion of a bond angle. For example, writing water as H-O-H does not say whether the angle between the two OH bonds is linear (180°), a right angle at 90°, or something in between. The evidence against a linear structure is extremely strong. Water is a very polar substance, meaning that it has a positive and a negative side. It has a very high dielectric constant, acting as a kind of shield against electric fields. These properties are completely inconsistent with a linear structure. Furthermore, water has, for its size, an abnormally high boiling point. Small molecules composed of light elements usually boil at low temperatures; water doesn't fit this pattern. The high boiling point suggests that water molecules attract each other in a moderately strong fashion, that in addition to the strong chemical bonds between the oxygen atom and its two hydrogens, there are also weaker but substantial bonds among water molecules as a whole. Much the same sort of thing can be said about ammonia, NH_3, which is also polar. The structure surely is not planar, but must be bent, somewhat like an umbrella, although, since ammonia does not have nearly as high a boiling point as water (-78° C versus 100° C for water), the weak bonds between ammonia molecules are not nearly as strong as those between water molecules. Alcohols also have high boiling points (methanol boils at 65°C, and is quite polar), which again suggests that the C-O-H bond is bent. The evidence indicates that compounds such as water and alcohols form weak "hydrogen bonds" between adjacent molecules, the positively charged H of one molecule attracting the negatively charged O of a nearby molecule, which explains why polar molecules tend to have elevated boiling points.

For single-bonded carbon, the evidence for a bent structure is much more subtle. For example, we can imagine two structures for methane, CH_4, a planar form in which the H-C-H bond angles are all 90°, and a tetrahedral form, which is harder to envision, in which all bond angles are 109°. Both structures would be non-polar, which agrees with experiment. However, consider the general compound CH_2XY, in which X and Y are two arbitrary functional groups (say CH_3 and OH, in which case the compound is ethyl alcohol). If the structure were planar, we could imagine two different compounds. In one, the X and Y would be on opposite sides; in the other, X and Y would be adjacent. But all

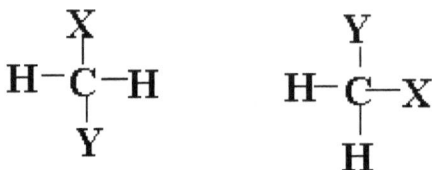

Figure 14.9 Hypothetical Planar Structures for CH₂XY

attempts to find two different structures for CH_2XY type compounds have failed, which strongly implies a tetrahedral model, in which the bond angle between any two groups attached to the central carbon is 109°, so that there is no such thing as "adjacent" or "opposite." (The reader will have to build models to be convinced of this; paper, being inherently two-dimensional, will not do the job.) An even more convincing case can be made when there are four different functional groups attached to the central carbon, as in CHXYZ. A planar structure would permit three different structures, with the H opposite X, Y or Z, while a tetrahedral structure allows for only (surprise!) just two. These two structures would be identical in every way but one: They would be mirror images of one another. This is essentially impossible to see on paper. To see the two forms properly, one **must** build models. Just as, when you or I raise our right hand in front of a mirror, our image raises his or her left hand, there are two forms of all CHXYZ compounds, right-handed and left-handed, mirror images of one another. This was first discovered by Pasteur, who observed that crystals of tartaric acid, $C_4H_6O_6$, found on the corks and sediment of wine bottles, come in two mirror-image forms, which he separated by hand using a pair of tweezers.

Not every bond angle in carbon compounds is 109°. When carbon forms a double bond, the bond angles are all roughly 120°, as in an equilateral triangle. Thus, when chemists draw a structure for ethylene, $CH_2=CH_2$, the model really is planar:

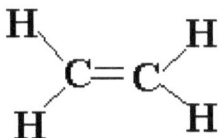

Figure 14.10 Structure of Ethylene

Furthermore, rotation about the central double bond does not occur; ethylene is rigid.

The evidence for a rigid, planar structure is again indirect and comes from compounds such as CHX=CHX, which exist in two forms, in which the two X groups are on the same ("cis") or opposite ("trans") side of the C=C axis. Once again it is the experimental presence (or absence) of different molecular forms that was the basis of chemists' conclusions.

Starting with simple rules and building upon them by studying (and synthesizing) hundreds and thousands of compounds, chemists were increasingly able to draw reasonably accurate (if schematic) three-dimensional structures of ever-larger molecules. It is an astonishing intellectual achievement. The most important chemical discovery is simply this: Three-dimensional molecular structure is the single most important factor in determining the real-world properties of chemical behavior. As we will see, life itself depends on molecular structure.

And what did the physicists of the time say about what the chemists were doing? Very little. The division between chemistry and physics was fairly sharp, and most physicists regarded the chemists' work with some suspicion. What were these molecules and atoms anyway? How big were they? Did they exist at all? What were they made of? Convenient hypotheses, certainly, but nothing much to do with physics or with physicists' sense of reality.

This attitude began to crumble at the beginning of the twentieth century, with J.J. Thompson's discovery of the electron, Rutherford's discovery of the nucleus, and Einstein's discussion of Brownian motion in 1905. Einstein theorized that if molecules existed (!), their size and motion could be observed in the random motion of dust particles in the air, easily visible in a shaft of sunlight. If molecules of air existed, and if they were of a certain minuscule size, some of them should be moving at very high speeds, and if at these speeds they bumped into dust particles, then the dust particles ought to dart about in a random but partially predictable fashion. Bingo! Physicists, at last, had a way of seeing the effects of molecules using mathematical tools they were accustomed to, and the sharp division between physics and chemistry began to blur. Indeed, it soon blurred so much that physicists were forced to invent an entirely new branch of their field -- quantum mechanics -- to explain atomic and molecular behavior, and quantum mechanics turned out to violate many of the rules of nineteenth-century physics, at least in so far as very small particles were concerned. Einstein was a co-inventor of the new field, although his main contribution to physics was in relativity theory, not quantum mechanics. Still, even after physics and chemistry slowly began to merge, the two disciplines looked at molecular modeling in very different ways, with physicists focusing on simple structures using purely physical techniques, while chemists continued to churn out models of increasing size and complexity, dimly aware that their methods and rules were not universally valid. This difference between the two fields continues to the present day, although chemists routinely use tools adopted from physics (e.g., nuclear magnetic resonance and x-ray crystallography) to help them formulate their models.

There is a famous example of these two ways of thinking that is now just sixty years old: Watson and Crick's discovery of the structure of DNA, and Rosalind Franklin's failure to discover it herself, although she was the one who performed the critical x-ray crystallographic experiments that led to the structure.

There were several key pieces of chemical information about DNA that were widely known. First, DNA consists of just four units, adenine, guanine, thymine, and cytosine, usually abbreviated A, G, T, and C. Second, while the proportion of these four units varies from sample to sample, the amount of A always equals the amount of T, and likewise the amount of C always equals G. Third, the chemical compositions (and structures!) of each of the four units were well known, as shown below

:

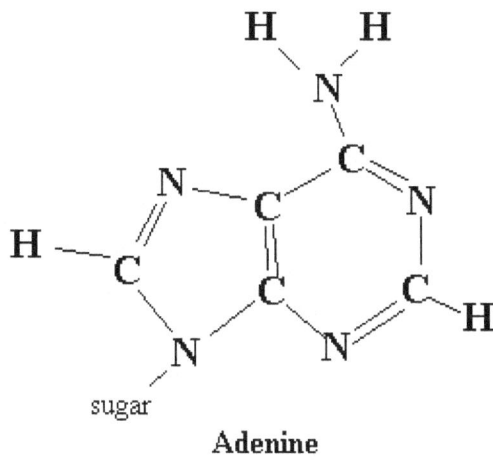

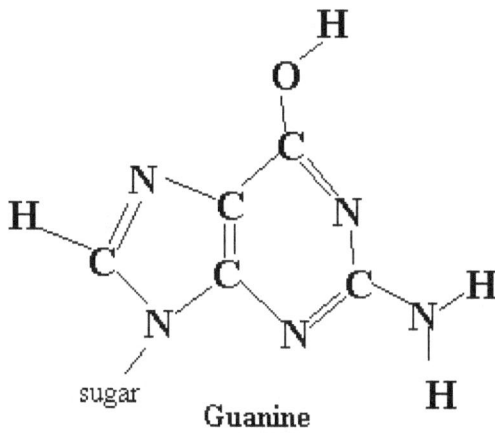

Figure 14.11 : Structures of Adenine and Guanine

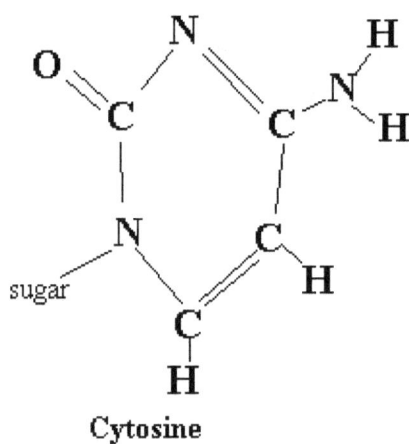

Cytosine

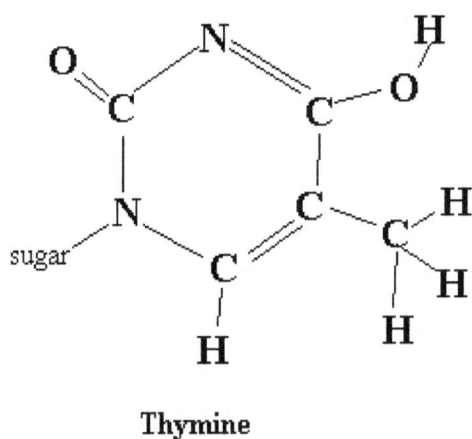

Thymine

Figure 14.12 Structures of Cytosine and Thymine

Fourth, from Rosalind Franklin's data, it was clear that DNA had some sort of helical structure, although what this meant was unclear since Linus Pauling had already shown that proteins -- an entirely different sort of biochemical -- could have a helical structure, and helices seemed merely a clever way for nature to package a large number of atoms and molecules into a small space. Fifth, Franklin's data showed that crystals of DNA belonged to the crystallographic space group monoclinic c2, meaning that the molecules had mirror symmetry, without left or right-handed forms, although this was a clue that everyone involved seemed to have missed.

In the tradition of chemists of the preceding one hundred years, Watson and Crick tried (in vain!) to build sensible models from the four units. They failed repeatedly until one of Jim Watson's office mates, Jerry Donohue, asked why Watson was using the enol forms for the structures of the bases. Because, Watson is said to have replied, that's what it said in the textbooks. Oh, said Donohue, that's wrong. Try the keto forms. These forms, illustrated below, are well-known alternatives to the structures written above, and we have already noted that acetone could in principle have an enol form.

The rest, as the saying goes, is history, remarkable history. Watson and Crick realized, almost immediately, that A and T could pair, through weak hydrogen bonds, the same sort of bonds that give water its high boiling point. Likewise, G and

C could pair through exactly the same sort of bonds, as shown below.

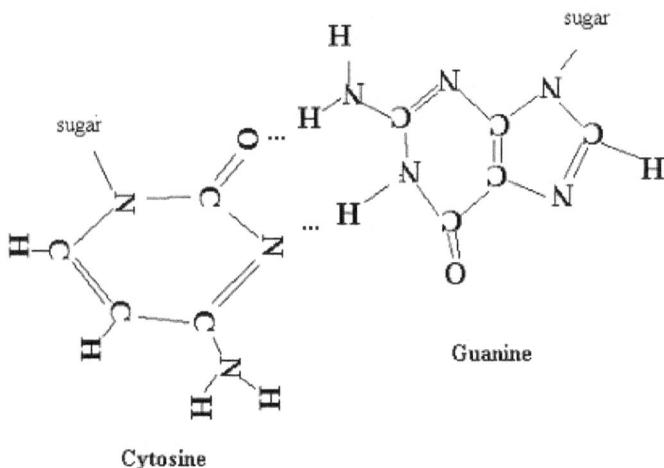

Figure 14.13 Pairing of Cytosine and Guanine via Hydrogen Bonds

This sort of structure was consistent with a helix, a double helix, in which the two chains are a kind of mirror image of each other. Furthermore, since the bonding between the two chains is through rather weak hydrogen bonds, the helix could be split apart into two separate chains or reformed into a double helix with little energetic input. The model fit all

224

the available facts, which were admittedly few, but it had the ring of truth to it: The model was too beautiful to be wrong. If nature didn't make DNA in this way, it should have!

It took years for Watson and Crick's model to be accepted. The evidence was skimpy although the model was plausible. In the end, the evidence accumulated, the model was confirmed.

For the past forty or fifty or sixty years people have speculated as to why Rosalind Franklin didn't get there first. After all, she was the one who knew the structure of DNA was a helix, quite likely a double helix. She knew that the basic repeat distance was 34 angstroms, the distance between successive turns of the helical chain. She knew that the A, T, G and C components were on the inside of the chain, with the phosphate groups (which are not shown here – they attach to the sugar molecules) on the outside. She alone had all the information needed to draw the proper structure. And she was exceptionally smart, capable, disciplined and well-organized: Her failure was hardly a matter of character flaws.

Many explanations have been put forth. She detested, for good reason, her work environment. As a woman, she felt belittled. She was on the verge of switching to another field in a new group at a new institution. Having worked for years in France, she found Englishmen boorish, intimidating and fatuous. All of these explanations are surely true. And yet, as Brenda Maddox's biography[28] shows, the central truth is that Franklin wasn't looking for a structure, or, if she was looking, her efforts were sporadic. She didn't think she had enough data to provide a structure. In a real sense, she wasn't in the hunt. Watson and Crick were. She thought like a physicist, which she wasn't (her doctoral degree was in physical chemistry). They thought like chemists (which they were not!). In the end, they were right (and she was wrong). The evidence, scanty and ambiguous as it was, was sufficient.

Problems and Exercises

Exercise #1: Draw a structure for ozone (O_3) that obeys the basic bonding rules.

Exercise #2: Find out (and draw) structures for the following molecules: (1) benzene; (2) cyclohexane; (3) glucose; (4) glycerine.

---

[28] Brenda Maddox, "Rosalind Franklin, the Dark Lady of DNA," HarperCollins, New York, 2002.

Exercise#3: How many different structures can be drawn for molecules of composition C_3H_5FO? (F stands for the fluorine atom, which forms just one bond.)

Chapter 15: Dating: How Old is the Earth?

A few years ago I felled a couple of very large spruce trees on my island property and counted the rings. Both were about 80 years of age, old enough for a spruce to die. One was still healthy, as far as I could tell; the other had begun to rot from the inside out. The healthy one was put to good use as spiling on the island pier; the other was consumed in the fireplace and the hot tub. I no longer remember why I cut them down. Perhaps they were leaning and ready to fall; perhaps they blocked a view of the sea. Both had passed the three score and ten years allotted to men and island spruce.

How do scientists know how old something is? Counting rings is tried and true, and tells us something about the climate for each year we count. There are trees that live far longer than man; some live for hundreds of years, and a few live for thousands. The bristlecone pine, found on the eastern edge of the Sierras, on the borders between California and Nevada, and between Nevada and Utah, is probably the longest-living species; one specimen was found to have over 4800 rings. Furthermore, while downed spruce rot quickly in the damp Maine maritime, fallen bristlecone pines scarcely rot at all in the arid east-slope climate of the Sierras. By comparing the ring patterns of living and dead trees, scientists have constructed a climate record that goes back over 9,000 years.

The bristlecone pine tells us more than just the climate record of the Sierra Nevada. It provides a way of calibrating the carbon-14 method of radioactive dating. Carbon consists of two isotopes, C-12, which comprises 98.89% of the total, and C-13, the other 1.11%. Both are stable. In addition, one carbon atom in 10^{12} is C-14, which is weakly radioactive, with a half-life of 5730 years. (C-14 emits an electron during decay.) As radioactive materials go, this is a short time. Even if carbon consisted solely of C-14 at some time in the past, its concentration would have dropped below a part in 10^{12} after a mere 200,000 years. Some process must be creating C-14. Indeed, at the top of the earth's atmosphere, cosmic ray neutrons collide with atmospheric nitrogen to produce C-14,

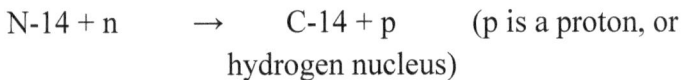

$$\text{N-14} + \text{n} \quad \longrightarrow \quad \text{C-14} + \text{p} \quad \text{(p is a proton, or}$$
$$\text{hydrogen nucleus)}$$

and the carbon reacts with oxygen to form carbon dioxide or CO_2. The CO_2 is taken up by plants and by the ocean; the plants are eaten by animals, and so on. We all have a little radioactive carbon in our bodies. The concentration of C-14 at any one place, at any one moment in time, is a balance between replenishment at the top of the atmosphere and decay, some of which occurs within our own bodies. When a plant or animal dies, exchange with C-14 in the atmosphere ceases, and the concentration of C-14 in the remains – wood, charcoal, bones or teeth - slowly decreases as the C-14 decays. Even before death, the exchange will stop in any part of the plant or animal that is no longer biologically active. The rings of trees are just such a material. The C-14 content of a ring reflects the C-14 content of the atmosphere when the ring was created and the time that has subsequently passed.

It is tempting to assume that the concentration of C-14 in the atmosphere has not changed over the past hundred thousand years. Then the age t of any biological material can be calculated from the formula

$$t = (\tau/\log_e (2)) \times \log_e (N_0/N)$$

where τ is the half-life, N is the number of atoms of C-14 remaining, and N_0 is the number of C-14 atoms at the time exchange with the atmosphere ceased. All we need to do is count the C-14 atoms.

How can we count atoms? The simplest trick is to use the radioactivity of C-14. Suppose we have a fresh sample of charcoal, pure carbon, that is, and for simplicity let's suppose the sample size is exactly 12 grams. As anyone who has ever taken a chemistry course should be able to remember, 12 grams of carbon contains 6.02×10^{23} atoms, of which 6.02×10^{11} should be C-14 atoms (one part in 10^{12}). How many of these atoms will decay in a minute? With the aid of a mathematical trick, we can use the formula above to provide the answer. The trick is that for x very very close to 1,

$$\log_e (x) = 1 - 1/x$$

where we are using natural logarithms (base e). Thus,

$$t = (\tau/\log_e (2)) \times (1 - N/N_0) = (\tau/\log_e (2) \times (N_0 - N)/N_0$$

We are interested in the number of decays, $N_0 - N$, so

$$N_0 - N = N_0 \times t \times \log_e (2)/\tau$$

232

Converting from years to minutes, this formula says there should be 138.6 decays per minute in our 12-gram sample. In an hour, 8313 decays.

These are wonderfully convenient numbers. The age of any sample of biological material can be obtained with a Geiger counter! When the technique was first applied to historically well-dated samples, such as wood from an Egyptian tomb, the agreement between the scientists and the historians was remarkable. The method worked!

Small discrepancies appeared as the method was refined. There were errors of several percent. Three percent may not sound like much, but in 4500 years that's an error of 135 years. Sometimes the historical record was more accurate. The source of these errors was gradually traced to the assumption that the concentration of C-14 in the atmosphere is constant over time. By measuring the C-14 content of well-dated tree rings like those of the bristlecone pine, it became possible to calibrate and correct the C-14 measurements, thus ensuring that the C-14 dating method can be applied accurately, at least for samples that are younger than the 9000-year-old bristlecone record.

The C-14 method of dating biological samples can be used, with diminished accuracy, on samples up to 30,000 or 40,000 years old. However, after 6 half-lives (34,380 years), only one part in 2^6 (1 part in 64) of the original C-14 remains. The rate of decay in our 12-gram sample drops from 8313 decays per hour to 99. More to the point, contamination by younger material can easily ruin attempts at accuracy.

One pleasant result of research over the past few decades is that large samples of material (e.g., 12 grams) are no longer needed; in fact, a few milligrams often suffice. We noted above that a 12-gram sample of (young) carbon contains 6 x 10^{11} atoms of C-14, yet we measure that concentration by looking at the decay of a few hundred or a few thousand atoms, an incredibly tiny fraction of the whole. A far better technique is to count all the atoms in the sample, one by one. This can actually be done in a mass spectrometer, a device that is at least vaguely similar to an old-fashioned cathode ray (television) tube. The material is first broken down into atoms and then accelerated by an electric field. The stream of atoms is then separated magnetically into individual streams of different mass. The masses of C-12, C-13, and C-14 are entirely different, so in principle, it is a simple matter to count the atoms of each species and find the ratio. The fact that the C-12 counts are at least 10^{12} times as great as the C-14 counts is no obstacle: Computers are good at counting things. A potential obstacle comes from nitrogen, N-14 to be specific. The exact atomic mass of C-14 is 14.00291, while the mass of N-14 is 14.00307. This is a really tiny difference. It cannot be overcome merely by trying to exclude nitrogen from the sample because all biological samples contain nitrogen, and in any case, we require an accuracy of better than one part in 10^{12}. Fortunately, the solution is remarkably simple. Nitrogen and carbon differ in their ability to form ions; carbon easily forms a negative ion, nitrogen does not. The two can be completely separated despite having nearly the same mass. Thus, using mass spectrometry, carbon samples as small as a few milligrams can now be dated accurately, if, of course, they are younger than the oldest bristlecone pine samples. If not, if the material is older, it can still be dated, to a limit, perhaps, of 75,000 years. If an unknown error of a few percent is still present, the achievement is still remarkable.

Similar methods, using radioactivity or mass spectrometry, can be used to date even older materials, such as rocks. However, the analysis is not quite the same as carbon-14 dating. Rocks do not live or die, nor are they much exposed to gamma rays. The starting point of time for a rock sample is the date on which it solidified from a melt. To understand this, I need to back up a trifle and delve a little way into the history of the universe, which began, as far as we know, with nothing but hydrogen and helium. Under the influence of gravity, stars formed and hydrogen atoms began colliding with other hydrogen atoms, eventually setting off thermonuclear reactions when densities and temperatures became great enough. This produced even more helium, but it did not really make much of any other element. The reasons are a little tricky to understand, but it turns out that making helium is not hard (for a star!), but generates so much heat that other "nucleosynthetic" processes are shut down. Heavier elements appeared on the scene only when big young stars grew old, used up their hydrogen, and collapsed. In the collapsed state, with its extremely high densities (and somewhat lower temperatures), heavier elements were created. The ensuing explosion spewed these heavier elements into space, where they were eventually captured by another newly forming star, and so on. Elements beyond iron can only be created in a supernova, where the densities were once so great that nuclei, protons, and neutrons, were once fused together in a gigantic unstable mass that created the really heavy elements such as germanium and bromine and molybdenum and silver and tin and platinum and gold and mercury and lead and bismuth and radium and uranium. These are all familiar elements that cannot be created from lighter elements by addition of a passing proton or neutron (or two) because they are unstable with respect to iron. Figure 15.1, the well-known "curve of binding energy[29]," illustrates why. Elements lighter than

---

[29] The mass of a single proton is 1.00797 "atomic mass units" (amu), which means that a mole of protons (6.02×10^{23} protons) has a mass of 1.00797 grams. Likewise the mass of a single neutron is 1.00866 amu. But the mass of a deuterium nucleus (one proton, one neutron) is 2.01410 amu, that is, 0.00253 amu less than the sum of its constituents (or 0.00127 amu less per nucleon.) Similarly, the mass of a helium nucleus (two protons, two neutrons) is 4.0026 amu, or 0.00767 amu lighter per nucleon. This difference is the binding energy. Iron, which has 26 protons and 30 neutrons, has the largest binding energy of any element,

iron, and elements heavier than iron, are not as stable[30] as iron. The gold in your jewelry came from a supernova many billions of years ago.

0.00950 amu. But uranium 238, with its 92 protons, has a binding energy of only 0.00818 amu.

[30] The fact that an element is in principle unstable with respect to iron does not mean that it will eventually decay to iron. By definition, only elements that are radioactive decay spontaneously. However, in the interior of a star, under the right conditions, the nuclei of different elements can interact to reach the most stable combinations.

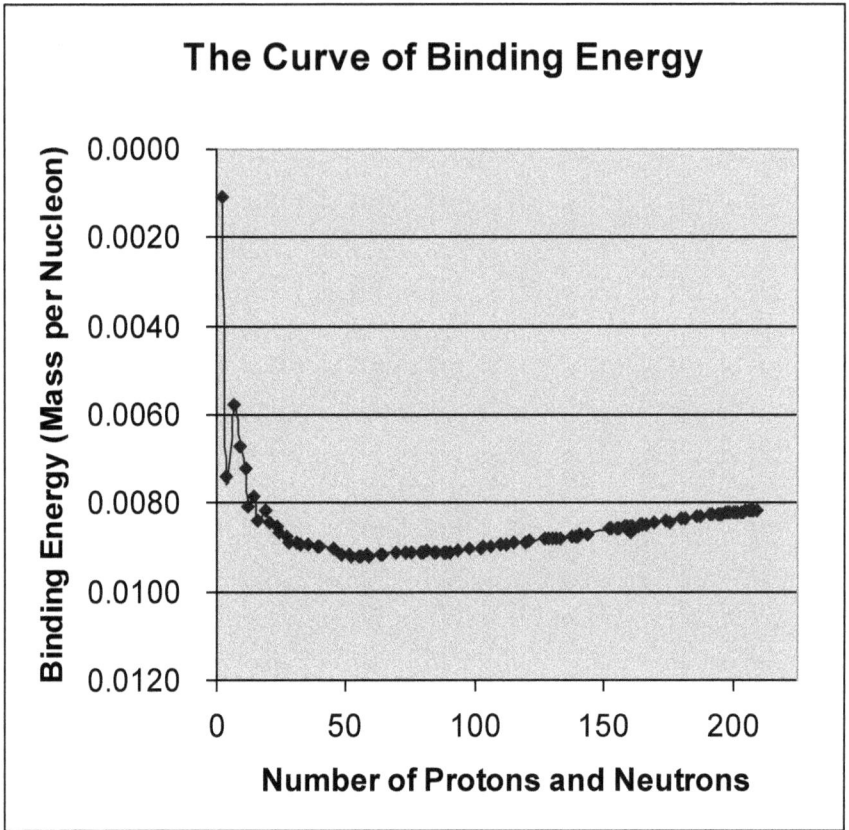

The Curve of Binding Energy

Figure 15.1 The Curve of Binding Energy: Mass Loss per Nucleon

Before the earth formed, the heavy, unstable elements had already begun to decay. Some decayed quickly, others slowly. Uranium-238 has a half-life of 4.5 billion years, which is why it has not yet disappeared; everything else that is heavier has a much shorter life, which is why these elements are no longer found.

When the earth formed and cooled, the elements began to separate. Iron mostly went to the core; lighter elements such as hydrogen and carbon and nitrogen and oxygen, stayed on the surface, at least in part. As materials solidified, they tended to purify and segregate. This is a natural phenomenon for solids that form from a melt. There is a tendency for impurities to be expelled when a substance crystallizes. Just as sea ice has much less salt than the waters from which it was formed, so any solidified melt tends to be purer than its original material.

In the history of our planet, rocks have melted and solidified many times over. In addition, rocks dissolve in water, form from animal shells, are crushed below layers of salt and silt, and are uplifted again in earthquakes and volcanic eruptions and plate tectonics. Geology, like any history, is messy and full of surprises. Comets and meteors crash into our planet, temperatures rise and fall, rocks melt and crystallize. The age of a rock is simply the time since its last formation from a melt. Many elements in a given rock may be radioactive and, from the composition of the rock and the concentrations of the radioactive byproducts, an age can be calculated. These procedures are subject to all kinds of errors. The process of crystallization may be rapid, trapping impurities, just as sea ice still contains some salt. Estimating the date of a rock is rarely simple. Nevertheless, there are all kinds of cross-checks. The results of one radioactive decay series can be compared to others. Independent estimates can be made from periodic marine depositions and tidal patterns. Dating rocks is not a simple matter, but when the pieces of a puzzle fit together in a way that defies alternative explanations, geologists can be pretty sure they have got the picture right.

One fairly simple nuclear decay scheme is that of potassium-40, usually written $_{19}{}^{40}K$. Potassium is a very common element, found in every rock and living creature. Most potassium (93.1%) is $_{19}{}^{39}K$, an isotope containing 19 protons and 20 neutrons in its nucleus. Some 6.9% of the element is $_{19}{}^{41}K$, with 22 neutrons. Both isotopes are stable. A very small proportion, 0.012%, is potassium-40 with 21 neutrons. This isotope decays to argon-40 with a half-life of 1.25 billion years:

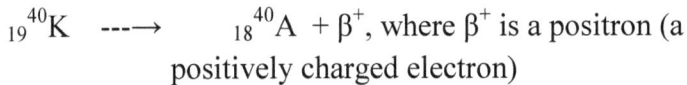

$$_{19}{}^{40}K \quad \text{---}\longrightarrow \quad _{18}{}^{40}A + \beta^+, \text{ where } \beta^+ \text{ is a positron (a positively charged electron)}$$

Argon is a gas, so it escapes readily from hot melted rock. However, once the rock solidifies, the argon is trapped, and its proportion to potassium-40 can be measured by mass spectrometry.

Our planet formed some 4.5 billion years ago, so say the oldest rocks found so far[31]. This is consistent with the views of the cosmologists, who, on the basis of entirely independent evidence, believe the universe itself is 13.5 billion years old. Our planet is not, by this reckoning, very old, and is scarcely middle-aged. Many stars lived and died before our local region of the universe began.

Problems and Exercises

Exercise #1: How many protons and neutrons are in the nucleus of an iron atom?

Exercise #2: What is "depleted" uranium? How is it made? To what extent is it radioactive?

Exercise #3: Plutonium is not found in nature. How is it made?

Exercise #4: What is deuterium? Is it radioactive?

Exercise #5: Tritium was used in some of the early digital wristwatches with LCD displays to provide nighttime visibility. What is tritium? How does it decay? What is its half-life?

---

[31] Because of plate tectonics, earth rocks this old are rare. However, meteor rocks and moon rocks have both been dated to 4.5 billion years ago.

Exercise #6: Potassium-40 and argon-40 have essentially the same masses, 39.96238 amu. How can mass spectrometry distinguish between these two species?

Exercise #7: Figure 15.1 is labeled "The Curve of Binding Energy," but the units of binding energy are given in mass per nucleon. Using Einstein's formula, $E = mc^2$, calculate the energy released when a mole of protons is fused with a mole of neutrons to form a mole of deuterium atoms.

About the Author

Peter O'D. Offenhartz received his bachelor's degree from Swarthmore College in 1960, where he majored in chemistry with minors in physics and mathematics. In 1963 he received his doctorate in physical chemistry from the University of Pennsylvania. Thanks to awards from the National Science Foundation and the National Institutes of Health he spent postdoctoral years at Cambridge University, the Institute for Solid State Physics (Tokyo), and Harvard University. He then taught chemistry at the University of Colorado (Boulder) and at Amherst College. Subsequently, he briefly worked as an author with the Physical Science Group under Uri Habersheim, and later joined EIC Laboratories as a Senior Scientist doing energy-related research. After leaving EIC he was involved in the startup of several independent computer companies, including Starbuck Data Company, Software Tailors, and MATTimekeeping Systems. He is the author of "Atomic and Molecular Orbital Theory (McGraw-Hill)," a graduate text, as well as many scientific papers. He is married to Barbara Hopf Offenhartz, has two grown children, John and Kathleen, and three grandchildren, and lives in Wellesley, Massachusetts. He and his wife spend their summers on Eagle Island, East Penobscot Bay, Maine.

www.ingramcontent.com/pod-product-compliance
Lightning Source LLC
Chambersburg PA
CBHW071630200326

41519CB00012BA/2239